Managing Complexity in MRT Renewals

emerald PUBLISHING
ice Publishing

Managing Complexity in MRT Renewals

Systems, Interfaces, Governance and Leadership in Live Urban Rail Networks

Daniel Woods
MSc, MBA, CEng

Published by Emerald Publishing Limited,
Floor 5, Northspring, 21-23 Wellington Street, Leeds LS1 4DL.

ICE Publishing is an imprint of Emerald Publishing Limited

Other ICE Publishing titles:
Designing and Engineering Economic Tramways
Lewis Lesley 978-1-83708-399-2
Cities for Driverless Vehicles
Xu Shang 978-0-7277-6452-2
Tunnelling in the Urban Environment
Lidija Zdravkovic 978-0-7277-6377-8

A catalogue record for this book is available from the British Library

ISBN 978-1-80791-579-7

Cover photo: shutterstock.com/g/PixelBiss

Commissioning Editor: Viktoria Hartl-Vida
Content Development Editor: Ryan Molyneux
Production Editor: Emma Sudderick
Project Manager: Krishna Prasad P

Typeset by: KnowledgeWorks Global Limited

Contents

Preface

Metro and mass rapid transit (MRT) networks are among the most complex pieces of infrastructure operated in modern cities. They are expected to run reliably every day, often with very limited tolerance for disruption, while carrying large passenger volumes through dense and constrained urban environments. This complexity is heightened in systems with multiple intersecting lines, shared stations and heavily used interchange points, where assets, operations and passenger movements are tightly interdependent. As these systems mature, renewal becomes unavoidable. Assets designed for long service lives must be replaced, upgraded or fundamentally altered while the railway continues to operate.

This book draws on experience gained within that reality. It reflects experience gained across long running renewal programmes in live metro and MRT environments, as well as related urban and mainline rail systems, across Europe, the Middle East and Asia, where technical decisions are inseparable from operational constraint, organisational context and leadership judgement. It is not a technical manual and it does not attempt to prescribe solutions. Instead, it articulates patterns that recur across renewal activity on urban rail networks and provides a structured way of thinking about why risk emerges, how complexity accumulates and where judgement matters most.

Metro and MRT renewals differ fundamentally from projects delivered largely outside live operation. They unfold over decades rather than years and are delivered within operating systems that cannot easily be paused or reset. Decisions are made under sustained pressure, with limited access, evolving technology and changing organisational conditions. In these environments, failure rarely arises from a single technical error. It more often emerges from interactions between assets, interfaces, people and governance arrangements that are poorly understood or insufficiently aligned.

The intent of this book is to support professionals who carry real accountability for safety, service continuity and public trust within metro and MRT systems. It is written for engineers, programme leaders, asset managers and decision makers who already understand the technical fundamentals, but who are seeking a clearer way to reason about renewal risk in complex, live railways. While the focus is on metro and MRT networks, many of the observations and principles described are transferable to other safety critical infrastructure systems operating under similar constraints.

This book does not criticise specific organisations, regions or delivery models. Metro and MRT systems operate in diverse regulatory, cultural and political environments, and each network faces its own constraints and trade-offs. The challenges described here are not presented as failures, but as predictable consequences of renewing ageing infrastructure within operating systems that are expected to remain available. The examples and patterns discussed reflect common conditions observed across many networks rather than the practices of any single system.

Throughout the book, emphasis is placed on system behaviour, interfaces and decision making under constraint. Particular attention is given to heavy civil renewals, access limitations, disruption planning and the governance arrangements that shape how renewal programmes unfold over time. These are the areas where technical compliance alone is rarely sufficient, and where engineering judgement, organisational capability and leadership consistency play a decisive role.

This book is intended to be read reflectively. Some readers may choose to focus on specific chapters relevant to their role or current challenges. Others may read it in full to build a broader system perspective. In either case, the aim is to provide a framework for thinking rather than a set of answers, and to encourage more deliberate engagement with the complexity inherent in renewing live metro and MRT networks.

If this book helps readers recognise familiar patterns, anticipate emerging risk earlier or approach difficult renewal decisions with greater clarity and confidence, it will have achieved its purpose.

All figures are original and created by the author unless otherwise stated.

Author Biography

Daniel Woods, MSc, MBA, CEng, is a Chartered Engineer and rail systems professional with over 23 years' experience delivering railway, mass rapid transit (MRT) and light rail transit (LRT) construction, renewal, maintenance and integration programmes across the UK, the Middle East and Southeast Asia.

Managing Complexity in MRT Renewals: Systems, Interfaces, Governance and Leadership in Live Urban Rail Networks

Daniel Woods
ISBN 978-1-80791-579-7
https://doi.org/10.1108/978-1-80791-578-020261001

Chapter 1
The nature of MRT renewal complexity

Mass rapid transit (MRT) networks are now core urban infrastructure in many cities, carrying large passenger volumes every day and operating under intense public and political scrutiny. As these systems mature, renewal becomes unavoidable. Assets that were designed for long life and low tolerance for failure must be renewed while the railway continues to operate, often with limited access and limited margin for recovery.

This chapter introduces the nature of MRT renewal complexity and explains why renewal behaves differently from standalone capital projects. It establishes the central argument of this book, that the most consequential risks are created by system interactions, interfaces and decision making under constraint. Later chapters build on this foundation by examining how MRT systems behave, where risk concentrates and why heavy civil renewals eventually challenge the assumption of continuous operation.

Across the world, MRT networks have expanded rapidly over recent decades, becoming essential infrastructure in cities of very different scale, density and governance. Systems operate in diverse regulatory, cultural and operating environments, yet many face similar long-term challenges as assets age and networks mature. Renewal programmes are now a defining feature of established MRT systems globally, regardless of geography or ownership model.

While each network has unique characteristics, the pressures associated with renewing assets in live, highly utilised railways are widely shared. The observations in this book are therefore not specific to any one city or organisation. They reflect recurring patterns seen across MRT systems internationally, and are intended to be applicable wherever complex rail networks must be renewed while maintaining public service.

1.1. Renewal in a live MRT system

MRT renewal is often described as the replacement or refurbishment of ageing assets. While technically accurate, this description does not reflect the reality of renewal within an operating urban railway.

MRT renewal takes place in a live system that continues to carry passengers while work is planned and delivered. Access is limited, operational margins are small, and the consequences of error are severe. The railway does not pause to accommodate engineering activity. Instead, renewal is forced to coexist with daily operations, public expectation and regulatory scrutiny.

In this environment, renewal is not a discrete technical task. It is an intervention into a system whose behaviour has evolved over decades. Assets, operating rules, maintenance practices and organisational structures are tightly linked. Any change made for renewal purposes alters system behaviour, often in ways that are difficult to predict in advance.

1.2. Why MRT renewals are fundamentally different

MRT renewals differ in important ways from standalone infrastructure projects.

Traditional project models assume a defined scope, a clear start and finish and a clean transition from construction to operations. Renewal activity rarely meets these conditions. The system remains operational throughout delivery. Scope evolves as new constraints emerge. Construction and operations overlap during testing, commissioning and staged handback.

Renewal programmes also extend over long periods of time. Leadership changes, standards evolve and delivery organisations are restructured. Decisions made early in the programme shape what is possible many years later, even when the original context no longer applies. Knowledge is lost as experienced staff move on, while accountability becomes distributed across multiple layers of governance.

Most importantly, renewal outcomes are realised incrementally. The railway passes through a series of intermediate operating states, each introducing new risks. These transitional states often present greater safety and performance challenges than either the original or final configuration. Failure in MRT renewals rarely reflects a single technical error. It more often emerges from cumulative system interactions that are poorly understood or weakly governed.

1.3. Complexity as an inherent feature of MRT renewal programmes

The difficulties encountered in renewal programmes are often attributed to poor planning or execution. While these factors contribute, they do not explain why similar problems arise repeatedly across different networks and jurisdictions.

Complexity in renewal work is inherent. It arises from the structure of the system itself.

MRT networks are composed of assets that are highly interdependent. Changes to track, power, signalling or structures rarely remain confined to a single discipline. Seemingly minor modifications can propagate through the system, affecting timetable robustness, maintenance effort, ride quality or safety margins.

At the same time, MRT systems operate with little tolerance for variability. Access windows are short. Recovery time is limited. Public tolerance for disruption is low. When variability occurs, it cannot easily be absorbed and instead reappears at interfaces between teams, contracts or operational functions.

No single individual or group has complete visibility of the system. Decisions are therefore made with partial information, under time pressure and within governance arrangements that may not align with how risk actually manifests.

1.4. The limits of planning and control

MRT renewal programmes rely heavily on planning, assurance and control processes. These are essential tools, but they have limits that must be understood.

Plans often assume stable access, predictable interfaces and consistent capability over time. In practice, access erodes, interfaces shift, and experienced personnel rotate out of key roles. Early decisions constrain later options, even when circumstances change. Information that appears complete during planning may prove inadequate during delivery.

A particular risk arises when compliance with process is mistaken for control of outcomes. Meeting assurance gates and procedural requirements can create confidence that risk is being managed, even as underlying system conditions degrade. Early warning signals that challenge the approved plan may be discounted or escalated too slowly.

Effective renewal depends on recognising uncertainty and responding to it, rather than assuming it can be fully designed out.

1.5. From project delivery to system stewardship

If system renewals in live operational networks are treated solely as projects, critical risks will continue to sit outside formal management structures. Renewal activity must instead be understood as a long-term system challenge.

This shift requires a different emphasis. Success depends less on optimising individual work packages and more on managing interactions between assets, operations, people and governance. Engineering judgement becomes as important as technical compliance. Decision making authority must sit close to where risk emerges, not simply where contractual responsibility lies.

System stewardship does not reduce the need for discipline. It increases it. It demands clarity of accountability, active management of interfaces, and leadership that is willing to challenge assumptions that no longer hold.

The chapters that follow build on this perspective. Chapter 2 examines the MRT network as a system, exploring how assets, operations and people interact to produce behaviours that shape renewal outcomes over time.

1.6. The coming challenge of heavy civil renewals

Much of the discussion around MRT renewal implicitly assumes that work can be accommodated within existing access arrangements. In practice, this often means night works or short engineering windows once revenue service has ceased. For many asset types, this model has proven workable, if increasingly constrained.

However, not all renewals will fit within this paradigm.

As MRT networks age, the need to renew heavy civil assets such as concrete slab track, drainage systems and underground structural elements becomes unavoidable. These assets are deeply embedded within the railway and are not easily segmented into small work scopes. Their renewal often involves extensive demolition, reconstruction and curing periods that cannot be compressed into short possessions without unacceptable risk.

When renewal activity exceeds what can be delivered within normal engineering hours, the problem changes fundamentally. The question is no longer how to optimise access, but whether continued operation during renewal is feasible at all. Extended closures, staged shutdowns or long-term service suspensions become part of the engineering conversation, even when they are organisationally and politically uncomfortable.

MRT organisations are typically resistant to sustained shutdowns, having been shaped by decades of operational continuity. Yet for certain classes of heavy civil renewal, such decisions may be unavoidable. At this point, renewal activity begins to change the operating system itself, not just its assets. When that point is reached, the challenge extends beyond engineering delivery into governance, authority and public accountability (Figure 1.1).

These scenarios represent some of the most demanding situations faced in MRT renewal programmes. They are where technical judgement, system understanding and leadership are tested most severely. Later chapters return to this issue in detail, examining how such renewals can be planned, governed and executed without undermining the long-term integrity of the network.

These dynamics extend beyond engineering delivery into governance, organisational capability and leadership judgement, themes developed in the chapters that follow.

Figure 1.1 MRT renewal as a long-duration system condition

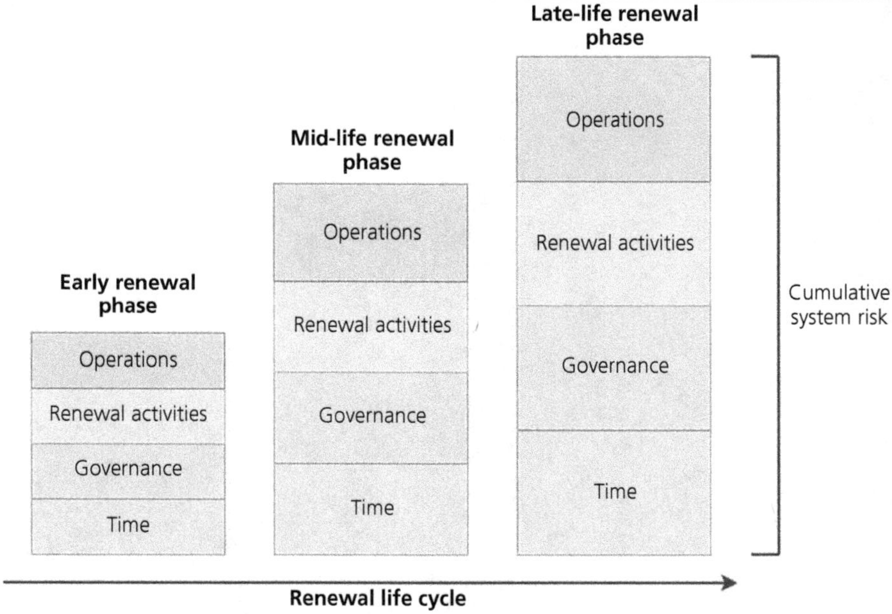

Managing Complexity in MRT Renewals: Systems, Interfaces, Governance and Leadership in Live Urban Rail Networks

Daniel Woods
ISBN 978-1-80791-579-7
https://doi.org/10.1108/978-1-80791-578-020261002

Chapter 2
Understanding the MRT as a system

This chapter looks at the MRT network as it actually operates, rather than as a collection of individual assets or discrete projects. It builds on Chapter 1 by exploring how assets, operations, people and organisational arrangements interact during renewal, and why those interactions often dominate risk.

In practice, many of the most serious renewal issues do not arise from technical failure within a single discipline. They arise from mismatches between system elements that were never intended to change at the same time. Understanding these interactions is critical to making sound engineering and management decisions. This chapter sets the context for later discussion on interfaces, access constraints and governance, where these effects become most visible.

2.1. The MRT is more than its assets

MRT networks are commonly described and managed in terms of assets. Track, power, signalling, rolling stock and structures are treated as distinct domains, each governed by its own standards, teams and renewal strategies. This framing is understandable. Assets are tangible, measurable and lend themselves to condition assessment, intervention planning and budget allocation.

However, an MRT does not behave as a collection of independent assets. It behaves as a system whose performance, safety and resilience emerge from the interaction between assets, operations, people and rules (Figure 2.1). Asset condition alone is rarely a reliable predictor of how the system will perform in service.

The MRT system includes far more than the core railway infrastructure. Station systems such as automatic fare collection, lifts, escalators, lighting, fire systems, building services and environmental control are integral to passenger safety, accessibility and network usability. These systems age, fail and require renewal in the same way as core railway assets. Their interaction with operations and passenger flows materially influences system performance.

However, not all assets influence renewal complexity in the same way. Many station and building systems can be isolated, zoned or replaced in stages, allowing renewal to proceed

Figure 2.1 MRT system elements and interactions

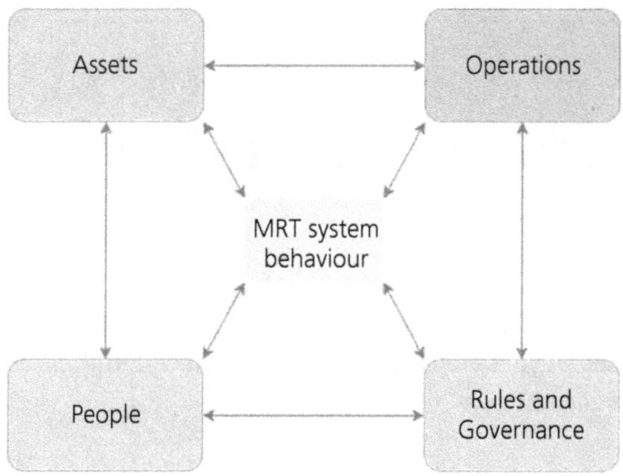

with limited impact on train operations. While their renewal can be logistically complex and operationally sensitive, it is often more tolerant of phased delivery and local disruption.

By contrast, core railway and heavy civil assets are deeply embedded within the system. Track, tunnels, drainage and structural elements are difficult to isolate and often constrain access, possession strategy and operational continuity. Renewal of these assets has a disproportionate influence on how the MRT system behaves, particularly during transition and recommissioning.

Understanding the MRT as a system therefore requires more than cataloguing assets. It requires judgement about which assets drive system behaviour and which drive renewal decisions. Treating all assets as equal from a renewal perspective obscures where risk truly concentrates.

2.2. Operational reality shapes system behaviour

The behaviour of an MRT system is shaped as much by how it is operated as by how it is designed.

Timetables, service frequencies, recovery margins and operating rules determine how tolerant the system is to disruption. Over time, these parameters evolve in response to demand growth, performance targets and public expectations. Margins are reduced. Services are intensified. Flexibility is traded for throughput.

These operational realities are not always fully visible during renewal planning. Designs are often developed against nominal assumptions that reflect how the railway was once intended to operate, rather than how it actually runs today. The result is a growing gap between technical intent and operational behaviour.

Renewal activity interacts directly with this operating context. Reduced margins mean there is little capacity to absorb variation during commissioning, degraded operation or early life defects. Issues that might previously have been manageable now propagate rapidly through the system. The same renewal intervention can have very different outcomes depending on the operating environment into which it is introduced.

Renewal programmes can unintentionally disrupt these mechanisms. Changes to assets, systems or organisational arrangements may remove long-standing adaptations without understanding the role they play. When this occurs, the system can become less stable even though assets are technically improved.

This is why the impact of renewal is often felt most acutely during transition periods. Assets may meet acceptance criteria, yet the system behaves unpredictably in service. The issue is not necessarily a technical failure, but a misalignment between renewed assets and the operational reality into which they are introduced.

During renewal work, understanding operational reality is central to managing risk. Engineering decisions that do not account for how the railway is actually operated will consistently underestimate system response once assets are returned to service.

2.3. People and competence as system components

People are not external to the MRT system. They are an integral part of how the system functions, adapts and remains stable over time.

In many ageing MRT networks, safe and reliable operation depends as much on experience and judgement as it does on formal procedures and asset condition. Drivers, controllers, maintainers and engineers develop a detailed understanding of how the railway behaves under normal and degraded conditions. This understanding is built over time through repeated exposure to faults, constraints and recovery scenarios.

Much of this knowledge is informal. It sits in habits, shared assumptions and established ways of working rather than in standards or documentation. While such practices are sometimes viewed as deviations from ideal operating models, they often act as stabilising mechanisms that allow the system to function despite asset degradation, design compromise or operational pressure.

Renewal activity directly affects these human elements of the system. Changes to assets, layouts or operating arrangements can invalidate long-standing assumptions about system behaviour. Organisational change, contractor interfaces and staff turnover can erode experience at precisely the point where it is most needed. When these effects coincide, the system can lose important stabilising influences without any obvious or immediate signal.

This risk is frequently underestimated. Renewal programmes often focus on technical assurance and procedural compliance, assuming that competence is transferable and interchangeable. In practice, the loss of local knowledge or experienced judgement can have a greater impact on system behaviour than the technical changes being introduced.

These effects are most visible during transition periods. Commissioning, staged handback and early operation following renewal place high demands on people. Decisions must often be made quickly, with incomplete information and limited recovery margins. In these conditions, experience and system understanding become critical safety controls.

In MRT networks that use advanced signalling and driverless operation, these dynamics can become more pronounced. Highly automated systems reduce reliance on direct human intervention during normal service, but increase dependence on tightly integrated assets, software and procedures. When renewal disrupts these systems, the role of people shifts rather than diminishes. Competence is required in different places, particularly during degraded operation, recovery and recommissioning, where system behaviour may be less predictable.

Recognising people and competence as system components has important implications for MRT renewal. It challenges assumptions about resourcing, organisational design and the timing of change. It also reinforces the need to consider how renewal activity affects not only assets and systems, but the human capability that allows the railway to function safely under pressure.

2.4. Interfaces as points of behaviour change

In MRT systems, risk rarely sits neatly within a single discipline, asset group or organisational boundary. It accumulates at interfaces (Figure 2.2).

Interfaces exist wherever responsibility, information or control passes from one part of the system to another. These may be technical interfaces between assets, operational interfaces between functions, organisational interfaces between teams, contractors or authorities. In steady state operation, many of these interfaces remain largely invisible. During renewal, they become active and consequential.

Renewal activity alters interfaces by changing assumptions that have often been stable for many years. Asset boundaries shift. Operating arrangements are modified. Roles and

Figure 2.2 Interfaces as points of risk concentration

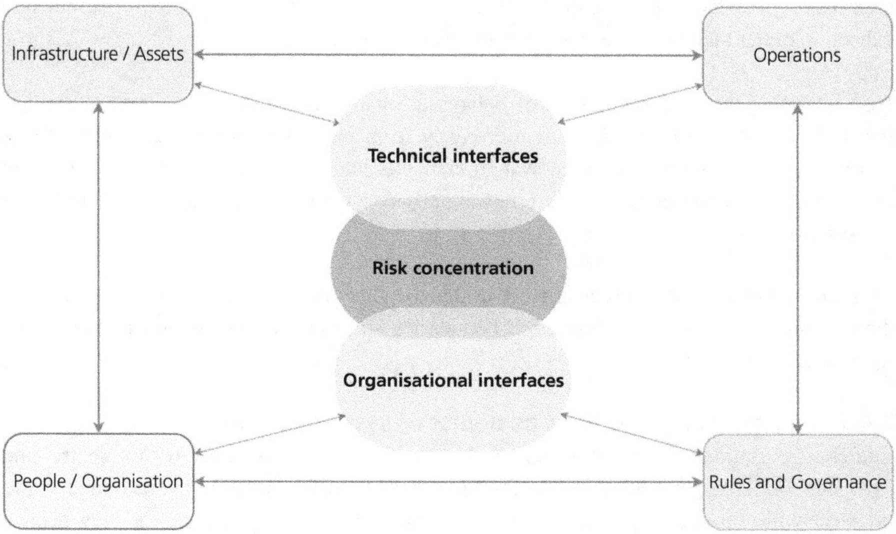

responsibilities are redefined. Each change introduces the potential for misunderstanding, delay or unintended interaction. While individual changes may appear manageable, their combined effect can alter how the system behaves as a whole.

Technical interfaces are a common source of difficulty. Renewed assets must integrate with legacy systems that were not designed to be modified. Tolerances, performance margins and failure modes may no longer align. Problems often emerge not because a component fails, but because two compliant components interact in an unexpected way once placed back into service.

Operational interfaces present similar challenges. Renewal frequently requires temporary operating arrangements, revised rules or altered response strategies. These changes rely on clear communication and shared understanding across functions. Where assumptions differ, the system can behave unpredictably, particularly during degraded or abnormal conditions.

Organisational interfaces are often the least visible and the most difficult to manage. Renewal programmes introduce new delivery structures, contractual relationships and assurance layers. Decision making authority may be distributed across multiple organisations, each with different priorities and incentives. In these conditions, accountability can become blurred and escalation paths unclear, even when formal governance arrangements exist.

A defining feature of interface related risk is that its effects are often displaced in time and location. Issues may only become apparent during commissioning or early operation, long after the original decisions were made. When problems arise, they are commonly attributed to execution rather than to the interface conditions that created them.

Understanding interfaces as points of behaviour change is therefore essential to managing MRT renewal risk. It shifts attention away from individual components or contracts and towards the boundaries where system elements interact. These boundaries are where uncertainty concentrates and where renewal activity most often challenges the stability of the system.

This understanding provides a critical bridge to Chapter 3, which examines interfaces directly and explores why unmanaged boundaries so often become failure points during renewal activity.

2.4.1 Interchange stations as points of system interaction

Interchange stations represent some of the most complex locations within metro and MRT networks. They are points where multiple lines, operating patterns and passenger flows intersect within a shared physical and operational environment. As a result, they behave less like discrete assets and more like system amplifiers, where local change can have wide reaching effects.

From a system perspective, interchange stations concentrate interfaces. Infrastructure, signalling, power, communications, station systems and operations overlap in ways that are often tightly coupled and historically evolved. Renewal activity at one interface can therefore influence multiple lines, timetables and operational regimes simultaneously. The effect of change is rarely contained within the immediate work area.

Passenger behaviour further increases complexity. Interchange stations are typically high demand nodes where dwell times, circulation capacity and crowd management are finely balanced. Even minor alterations to layout, access or service patterns can alter passenger flows across the wider network, affecting reliability and recovery well beyond the station boundary.

Renewal works at interchange stations also challenge organisational assumptions. Responsibilities are often distributed across multiple delivery teams, asset owners and operational groups, each aligned to a particular line or system. Without deliberate coordination, this fragmentation can obscure system level risk and lead to decisions that are locally rational but destabilising at the network level.

These characteristics make interchange stations particularly sensitive during renewal programmes. They require early system level thinking, explicit interface management and careful consideration of how temporary conditions will interact with normal operation. Treating interchange stations as isolated locations rather than as coupled system nodes significantly underestimates their influence on network behaviour.

Recognising interchange stations as system amplifiers reinforces the need to approach MRT renewal as a system problem rather than a collection of individual asset interventions. It also highlights why interface management, governance and operational insight are critical in locations where complexity is inherently concentrated.

During renewal work, problems often emerge in places and at times that were not anticipated. Issues may appear far from the location of the original work, or only after assets have been returned to service. This behaviour is frequently interpreted as poor execution or inadequate testing. In reality, it is often a natural consequence of how system elements interact under constraint.

Renewal activity introduces change into a system that is already tightly coupled. Assets are modified, operating arrangements are adjusted, people are asked to adapt and interfaces are disturbed. Each change may be justified and manageable in isolation. When combined, they can alter system behaviour in ways that are difficult to predict in advance.

One reason for this is that MRT systems operate with a limited margin. Performance, capacity and resilience are closely balanced. Small changes that would be absorbed in a less constrained environment can have disproportionate effects. A minor shift in asset behaviour, operating response or information flow can propagate rapidly through the system, particularly during peak operation or degraded conditions.

Another factor is the displacement of cause and effect. Decisions made early in a renewal programme may not reveal their consequences until commissioning or early operation. When issues arise, attention naturally focuses on the point of failure rather than on the sequence of interactions that created it. This makes learning difficult and encourages repeated mistakes across programmes.

The role of people further complicates this picture. As discussed earlier in this chapter, experience and informal knowledge often act as stabilising influences. When renewal activity disrupts these elements, the system may lose resilience at precisely the moment when it is under greatest stress. This loss may not be immediately visible, particularly if technical acceptance criteria have been met.

Highly integrated and automated systems can amplify these effects. Tight coupling between assets, software and procedures reduces tolerance for variation. When renewal changes assumptions embedded within these systems, behaviour can shift quickly and with limited opportunity for intervention. The resulting issues are often interpreted as system failures rather than as indicators of deeper interaction effects.

Understanding these dynamics is critical for MRT renewal. Unexpected system behaviour is not an anomaly to be eliminated through better planning alone. It is an inherent feature of renewal within complex, constrained systems. Recognising this allows organisations to focus less on assigning blame and more on anticipating where interaction effects are likely to emerge.

This perspective reinforces the need to pay close attention to interfaces. It is at interfaces that assumptions diverge, information degrades and responsibility becomes unclear. Chapter 3 examines these boundaries directly and explores why they are the primary locations where renewal risk manifests in MRT systems. This system perspective underpins the discussion of interfaces, access, governance and leadership that follows.

Daniel Woods
ISBN 978-1-80791-579-7
https://doi.org/10.1108/978-1-80791-578-020261004

Chapter 3
Interfaces: Where risk actually lives

MRT renewals rarely fail because a single asset or discipline performs poorly. They fail because boundaries are misunderstood, unmanaged or assumed to be stable when they are not.

This chapter focuses on interfaces as the primary locations where renewal risk accumulates. It builds on the system perspective established in Chapter 2 by examining how interfaces behave under change, why they are difficult to govern and how they become the points at which otherwise sound engineering decisions can lead to unexpected outcomes. Understanding interfaces is essential to managing renewal in live, constrained MRT environments.

3.1. What is an interface in an MRT renewal context

In the context of MRT renewals, an interface is any point at which responsibility, information, control or assumption passes from one part of the system to another.

Some interfaces are obvious and physical in nature. They exist between assets, such as the connection between track and signalling, or between power systems and rolling stock. Others are less visible. They exist between organisations, between operational functions or between different phases of work such as design, construction, testing and operation.

Interfaces are often treated as coordination issues rather than sources of risk. They are documented through interface registers, responsibility matrices or contractual provisions. While these tools are necessary, they tend to describe interfaces as static boundaries rather than as active parts of the system.

During renewal, interfaces rarely remain stable. Assumptions that were once valid begin to shift. Assets are modified. Operating arrangements are adjusted. Roles and responsibilities are redefined. Each change introduces the potential for misunderstanding, delay or unintended interaction. While individual changes may appear manageable in isolation, their combined effect can alter how the system behaves as a whole.

Many interface related failures do not arise because responsibilities were undefined. They arise because different parties held different assumptions about how the interface would

Figure 3.1 Types of interfaces in MRT renewals

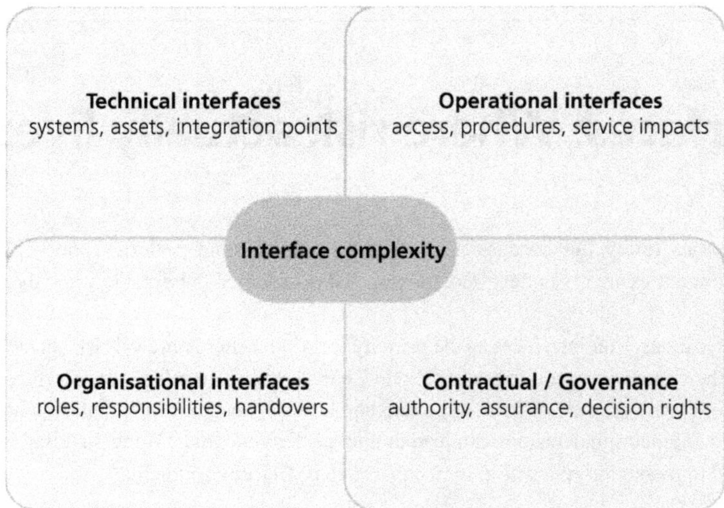

behave once the system was changed. These assumptions often remain untested until the system is placed back into service.

A further complication is that interfaces rarely exist in isolation. Changes at one interface can affect conditions at others, sometimes in subtle ways. This interconnectedness means that managing interfaces individually is rarely sufficient. What matters is how interfaces interact across the system during renewal activity.

Recognising interfaces as active elements of system behaviour is a necessary shift for renewal programmes. It moves attention away from compliance with interface documentation and towards understanding how boundaries actually function under pressure. This shift underpins the discussion in the sections that follow, which examine different types of interfaces and why they so often become the focal points for renewal risk (Figure 3.1).

3.2. Technical and physical interfaces

Technical and physical interfaces sit at the boundary between assets that must function together to support safe and reliable operation. In MRT renewals, these interfaces are often assumed to be well understood because they are grounded in engineering design and standards. In practice, they are among the most persistent sources of renewal risk.

Many MRT assets were not designed to be modified in isolation. Track, signalling, power, structures and rolling stock were developed as integrated systems, often based on assumptions that are no longer explicit or even known. Renewal activity disrupts these assumptions by introducing new components, new performance characteristics or new tolerances into an existing environment.

Problems at technical interfaces rarely present as outright failures. More commonly, they appear as marginal behaviour. Performance may degrade under certain operating conditions. Fault rates may increase without an obvious cause. Recovery times may lengthen. These effects can be difficult to trace back to a single interface change, particularly when each renewed element meets its individual acceptance criteria.

A common challenge arises where renewed assets must interact with legacy systems. Design standards evolve, but the surrounding system may not. Differences in stiffness, response time, control logic or failure modes can introduce subtle incompatibilities. These incompatibilities often remain latent until the system is placed under operational stress, such as during peak service or degraded operation.

Physical interfaces associated with heavy civil assets are particularly unforgiving. Interfaces between track and structure, drainage and formation or tunnels and fixed equipment are deeply embedded and difficult to access once work is complete. Errors or assumptions at these boundaries are costly to correct and may constrain future intervention options. As discussed earlier in this book, these interfaces frequently drive access strategy and renewal sequencing.

Another source of difficulty is the accumulation of change over time. Individual renewals may be small and well controlled. Taken together, they can progressively alter how forces, loads or signals propagate through the system. The original system behaviour may no longer apply, even though no single change appears significant in isolation.

Managing technical and physical interfaces therefore requires more than compliance with interface specifications. It requires an understanding of how renewed assets will behave when combined with the rest of the system, and how those interactions may evolve over time. Where this understanding is absent, risk tends to emerge late, often during commissioning or early operation, when options for mitigation are limited.

These characteristics explain why technical interface issues are frequently discovered after substantial effort has been invested and expectations have been set. They also explain why such issues are often mischaracterised as execution problems rather than as systemic interaction effects.

Recognising the dynamic nature of technical and physical interfaces is essential for MRT renewals. It shifts attention away from isolated design assurance and towards the behaviour of the system as a whole. This perspective is necessary before considering the operational and organisational interfaces that further shape renewal outcomes.

3.3. Operational interfaces

Operational interfaces exist where renewal activity meets the rules, practices and decisions that govern how the MRT is actually run. These interfaces are often more fragile than technical ones, because they depend on shared understanding rather than physical connection.

During renewal, operational interfaces are frequently disturbed. Temporary operating arrangements are introduced. Rules are revised. Responsibilities shift between engineering, operations and maintenance. Each change alters how information flows and how decisions are made in real time.

A common source of risk arises during transitions between construction and operation. Testing, commissioning and staged handback require close coordination across functions that do not normally operate together. Assumptions about readiness, fault response or degraded operation may differ between teams. These differences often remain hidden until the system is placed under live service conditions.

Operational interfaces are particularly sensitive to timing and pressure. Decisions that are manageable during planned activities become more difficult when services are running, recovery margins are limited and public scrutiny is high. In these conditions, even small ambiguities in authority or procedure can lead to hesitation, delay or inconsistent response.

Renewal also introduces new failure modes that are not fully captured in existing operating rules. Assets may behave differently during degradation. Fault patterns may change. Recovery strategies that were effective before renewal may no longer apply. Where operating procedures lag behind system change, staff are forced to rely on judgement under pressure.

These challenges are compounded when operational changes are treated as temporary. Temporary rules and arrangements often persist longer than intended, becoming normalised without full review. Over time, this can create a layered operating environment in which it is unclear which assumptions still hold and which have been superseded.

Managing operational interfaces during renewal programmes therefore requires more than procedural updates. It requires sustained attention to how the system is actually being operated, how decisions are made under pressure and how responsibilities are understood across functions. Where this attention is lacking, risk tends to surface during live operation rather than during planned engineering activity.

Figure 3.2 Interface failure propagation

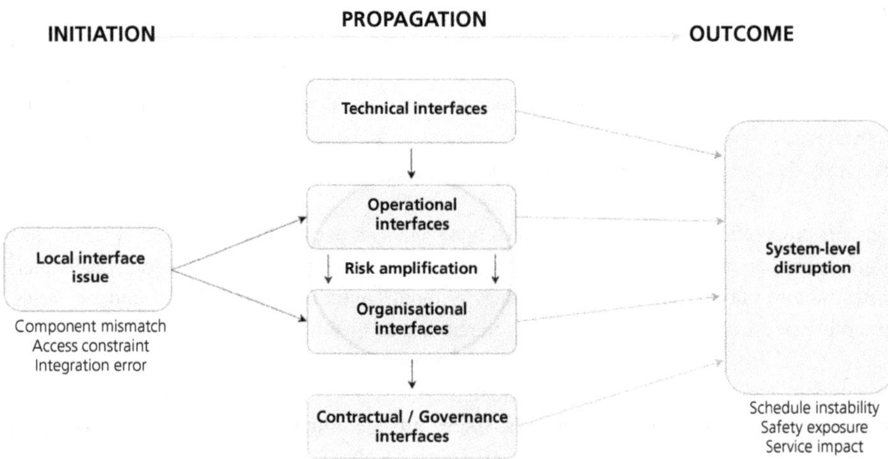

Operational interfaces sit at the intersection of assets, people and rules. They are where technical change meets operational reality. Understanding their behaviour is essential before considering the organisational and contractual interfaces that further shape renewal outcomes (Figure 3.2).

3.4. Organisational and contractual interfaces

Organisational and contractual interfaces define how authority, accountability and decision making are distributed across the MRT system during renewal. These interfaces are rarely physical, but they exert a powerful influence on how risk is recognised, escalated and addressed.

Renewal programmes often involve multiple organisations operating under different mandates. Asset owners, operators, delivery organisations, designers, contractors and assurance bodies each play a role. While formal governance structures may be clearly documented, the practical reality of who decides what, and when, is often less clear.

A common source of difficulty arises where organisational boundaries do not align with system risk. Technical responsibility may sit with one party, operational consequence with another, and authority to intervene with a third. When issues emerge at interfaces, this misalignment can delay action, dilute ownership or discourage escalation.

Contractual arrangements can reinforce these effects. Contracts are typically structured to define scope, responsibility and liability, but they are less effective at managing uncertainty or emergent system behaviour. When renewal activity produces outcomes that sit outside anticipated scenarios, contractual boundaries can become barriers to collaboration rather than mechanisms for resolution.

These challenges are often amplified during transition periods. Commissioning, staged hand-back and early operation require rapid decision making across organisational lines. Where contractual or governance arrangements emphasise separation over integration, the system can struggle to respond coherently under pressure.

Organisational interfaces are also affected by change over time. Sustained renewal programmes experience shifts in leadership, strategy and delivery models. Decisions made under one structure may be inherited by another. Without continuity of intent and understanding, assumptions embedded in earlier phases can be lost or misinterpreted.

The impact of organisational and contractual interfaces is rarely visible in isolation. It becomes apparent through patterns of behaviour. Issues are identified late. Decisions are deferred. Responsibilities are disputed. In these conditions, technical problems are often addressed symptomatically, while underlying interface issues persist.

Recognising organisational and contractual interfaces as active elements of the MRT system is essential for effective renewal. It highlights the need to align authority with risk, to support collaboration across boundaries and to maintain clarity of decision making throughout long-term programmes. Without this alignment, even well-engineered renewals can be undermined by fragmented governance.

Across extended renewal programmes, interfaces are where system behaviour is most sensitive to change. Technical, operational and organisational boundaries each shape how risk emerges, but it is their interaction that most often determines outcomes. When these interfaces are treated as static or secondary, renewal activity tends to expose weaknesses only once the system is placed back into service.

Recognising interfaces as active and dynamic elements of the MRT system shifts attention away from isolated failures and towards the conditions that allow risk to accumulate. This perspective is essential when considering access, possession and delivery constraints, where the consequences of interface misalignment become most acute. Chapter 4 turns to these constraints directly, examining how limited access shapes renewal strategy and compounds system risk.

These interface dynamics become most visible under access constraint, where delivery pressure forces assumptions into operational reality.

Managing Complexity in MRT Renewals: Systems, Interfaces, Governance and Leadership in Live Urban Rail Networks

emerald PUBLISHING ice Publishing

Daniel Woods
ISBN 978-1-80791-579-7
https://doi.org/10.1108/978-1-80791-578-020261005

Chapter 4
Access, possession and the reality of constraints

Access to the railway is the defining constraint in MRT renewals. It shapes what can be delivered, how risk is managed and where pressure accumulates across the system. While access is often treated as a planning input, in practice it functions as a governing condition that influences engineering decisions, organisational behaviour and system outcomes.

This chapter examines access and possession not as logistical challenges to be optimised, but as structural realities that frame all renewal activity in live MRT networks. It explores how limited access interacts with system behaviour, why delivery pressure compounds over time and how decisions made under access constraint often determine renewal risk long before work begins.

4.1. Access as a structural constraint

In most MRT networks, renewal work is delivered within tightly constrained access windows. These windows are typically defined by the end of revenue service and the start of operations the following day. Within this period, work must be set up, executed, inspected and handed back safely.

This access model is deeply embedded in how MRT organisations plan and deliver renewal. It shapes expectations about what is feasible and normalises the idea that complex engineering activity can be reliably delivered in short, repetitive windows. Over time, this becomes an assumption rather than a choice.

However, access is not simply a scheduling constraint. It is a structural condition that governs system behaviour. Limited access restricts the ability to inspect assets directly, increases reliance on preparation and assumptions and compresses decision making into narrow timeframes. It also reduces the margin available for recovery when work does not proceed as planned.

The cumulative effect of repeated short access windows is often underestimated. Fatigue builds across teams. Temporary arrangements persist longer than intended. Deferred work accumulates. Small deviations that would be corrected under longer access regimes are instead carried forward. Over time, these effects alter how the system behaves and how risk is distributed.

Access constraints also shape engineering decisions in subtle ways. Designs may be simplified to fit available windows. Interfaces may be deferred or broken into stages. Testing and commissioning activities may be compressed or resequenced. Each adaptation may be reasonable in isolation. Taken together, they can increase system fragility.

Importantly, access constraints are not neutral. They interact with organisational incentives and delivery pressure. Programmes are planned around possession availability. Success is measured by access utilisation. Failure is often defined as overrunning a possession rather than as introducing long-term risk. In this environment, decisions tend to favour short-term recoverability over long-term system integrity.

Recognising access as a structural constraint rather than a planning inconvenience is critical for MRT renewals. It reframes the question from how to fit work into available windows to whether the access model itself is appropriate for the class of renewal being undertaken. This distinction becomes increasingly important as networks age and renewal activity moves beyond routine intervention.

Over time, access regimes become cultural as well as operational, shaping organisational expectations about what is considered feasible or normal.

4.2. Night works and the illusion of repeatability

Night works form the backbone of renewal delivery in most MRT networks. They are familiar, structured and appear predictable. Over time, they become the default solution for delivering increasingly complex interventions within a live operating system.

The apparent strength of night works lies in their repeatability. The same access windows occur each day. The same activities are planned, resourced and executed in sequence. This creates confidence that complex work can be decomposed into manageable tasks and delivered reliably over time.

However, this confidence often rests on an assumption that conditions remain sufficiently stable from one night to the next. In practice, this assumption rarely holds.

Each night begins with a different starting condition. Asset state evolves. Temporary works accumulate. Interfaces shift. Weather, fatigue and minor delays compound. Small variations that appear insignificant in isolation begin to interact across successive possessions.

As renewal progresses, the system becomes increasingly sensitive to these variations. Activities that were achievable early in a programme become harder to complete. Recovery margins shrink. Dependencies increase. The window for error narrows.

Night works also encourage a form of temporal fragmentation. Work is planned to stop and start repeatedly, often leaving assets in intermediate states for extended periods. These intermediate conditions may never have been intended to exist in service or under maintenance. While managed carefully, they represent a persistent source of risk that grows as programmes extend.

Another consequence of night work reliance is the deferral of integration. Interfaces are often left until late in the programme, when multiple strands of work converge within limited access. By the time integration occurs, options for rework or adjustment are constrained. Issues that might have been resolved under longer access regimes instead surface during commissioning or early operation.

The illusion of repeatability is reinforced by planning artefacts. Schedules, possession plans and productivity metrics suggest stability and control. In reality, they mask a gradual erosion of margin and resilience. When the system eventually reacts, it often does so abruptly.

Understanding the limits of night works is essential for MRT renewals. Repeatability is not a substitute for robustness. Where the access strategy is misaligned with the nature of the work, risk accumulates quietly until it can no longer be absorbed within the available windows (Figure 4.2).

4.3. The compounding effect of delivery pressure

Access constraints do not exist in isolation. Over time, they interact with programme commitments, organisational expectations and public performance targets to create sustained delivery pressure.

Figure 4.2 Access windows against work scope

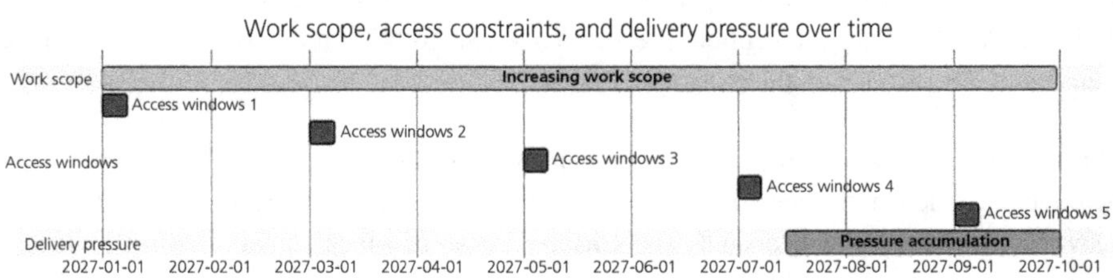

In long running renewal programmes, early success within limited access windows often sets expectations for what is achievable. Plans are built around assumed productivity. Commitments are made on the basis of past performance. As work progresses, these assumptions become embedded, even as conditions change.

Delivery pressure increases as programmes advance. Interfaces multiply. Dependencies tighten. Margins that were available early in the programme are gradually consumed. At the same time, tolerance for disruption typically reduces. Service demand grows. Public scrutiny intensifies. Opportunities to recover lost ground diminish.

Under these conditions, access windows take on increased symbolic importance. Completing work within the available possession becomes a measure of success in its own right. Decisions are framed around avoiding overruns rather than around managing long-term system risk. When difficulties arise, effort is directed towards maintaining the access plan rather than questioning whether the plan remains appropriate.

This pressure shapes behaviour across the system. Work is broken into smaller increments. Temporary arrangements are extended. Integration activities are deferred. Issues that cannot be resolved within the available window are carried forward. Individually, these decisions may appear reasonable. Collectively, they increase system fragility.

The effect is particularly pronounced in constrained environments such as underground networks. Logistics, access routes, ventilation and emergency arrangements limit what can be achieved within short periods. As renewal activity intensifies, the gap between what the work requires and what the access model allows becomes harder to ignore.

At this point, the system often exhibits warning signs. Fatigue increases. Temporary conditions persist. Recovery from incidents becomes more difficult. The railway remains operational, but with diminishing resilience. These signs are frequently interpreted as execution challenges rather than as indicators that the access regime itself is under strain.

Recognising the compounding effect of delivery pressure is critical. It highlights that access constraints are not static. Their impact grows over time as commitments accumulate and flexibility erodes. Without intervention, pressure continues to build until risk can no longer be managed within normal engineering hours.

This realisation marks a transition point in many MRT renewal programmes. It forces a reassessment of whether the prevailing access model remains fit for purpose, particularly for heavy civil and deeply embedded assets. Chapter 5 examines these situations directly, focusing on renewal activities that challenge the assumption of continuous operation and require fundamentally different approaches to access and possession.

Managing Complexity in MRT Renewals: Systems, Interfaces, Governance and Leadership in Live Urban Rail Networks

Daniel Woods
ISBN 978-1-80791-579-7
https://doi.org/10.1108/978-1-80791-578-020261006

Chapter 5
Heavy civil works in live MRT networks

Some renewal activities challenge the fundamental assumption that an MRT can remain continuously operational while work is undertaken. Heavy civil works fall squarely into this category.

This chapter focuses on renewal activities that are deeply embedded within the railway and cannot be easily segmented or deferred. Tunnel renewals, slab track replacement, drainage reconstruction and structural interventions alter the physical fabric of the system in ways that interact directly with access, safety and operational continuity. These works expose the limits of conventional delivery models and force decisions that extend beyond engineering execution into governance and leadership.

5.1. Heavy civil renewal as a system critical event

Heavy civil renewal is often planned and governed using the same structures applied to more routine interventions. This approach underestimates its impact.

Unlike component level renewal, heavy civil works affect large portions of the system simultaneously. They alter load paths, clearances, drainage behaviour and emergency arrangements. In tunnel environments, they also affect ventilation, access routes and evacuation assumptions. These effects cannot be isolated to a single asset or discipline.

The defining characteristic of heavy civil renewal is embedment. Slab track, tunnel linings and drainage systems are integrated into the surrounding structure and are difficult to access once work has started. Temporary conditions introduced during renewal may persist for extended periods and can themselves become sources of risk.

These works also introduce significant logistical constraints. Material delivery, spoil removal, workforce movement and emergency access are all limited by the physical environment. In underground networks, these constraints often dominate the delivery challenge. Productivity is governed less by engineering capability than by what the environment allows.

Figure 5.1 Heavy civil renewal within live operation

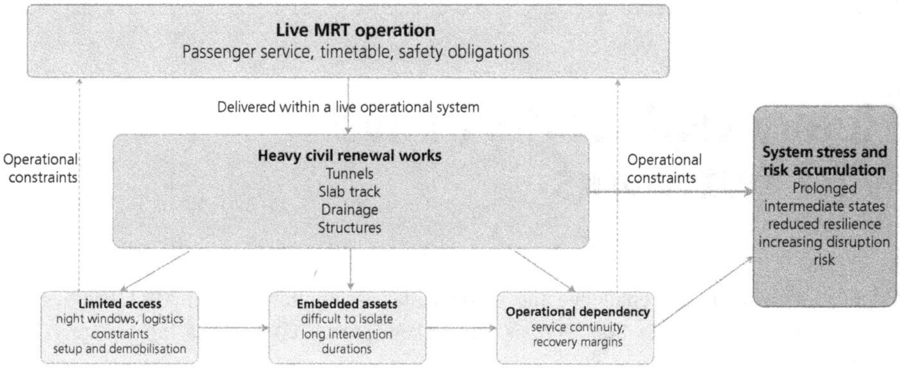

When heavy civil works are attempted within short access windows, the system is forced into prolonged intermediate states. Interfaces remain open. Temporary arrangements persist. Recovery margins erode. The railway continues to operate, but with increasing dependence on control and compliance rather than inherent resilience.

For these reasons, heavy civil renewals should be treated as system critical events. They require different assumptions about access, risk ownership and decision making. Planning them as an extension of routine renewal activity obscures their true impact and delays necessary choices about delivery strategy.

Recognising heavy civil renewal in this way does not imply failure or poor foresight. It reflects the reality of ageing MRT infrastructure. As networks mature, the work required to sustain them inevitably becomes more intrusive and more consequential. The question is not whether such renewals will occur, but how they are anticipated, governed and executed. At this point, engineering feasibility gives way to governance choice, as decisions about access and disruption become inseparable from leadership accountability (Figure 5.1).

5.2. Tunnel environments and logistical constraint

Tunnel environments impose constraints that fundamentally shape how heavy civil renewal can be delivered. These constraints are physical, operational and human, and they cannot be removed through planning alone.

Access to underground assets is limited by geometry, ventilation, emergency egress and available work space. Movement of people, materials and equipment is slow and tightly controlled.

Spoil removal competes directly with installation activity. Each action consumes time and capacity within a confined environment.

These constraints create a delivery regime that is dominated by logistics rather than by technical complexity. The feasibility of work is often determined not by what needs to be done, but by what can physically be moved in and out of the tunnel within the available time. Productivity assumptions that are reasonable in surface environments often prove optimistic underground.

Emergency arrangements further limit flexibility. Fire safety, evacuation routes and rescue access must be maintained at all times, including during renewal activity. Temporary changes introduced to facilitate work must be carefully controlled and frequently reviewed. In some cases, these arrangements become the critical path rather than the engineering task itself.

The interaction between tunnel logistics and access windows is particularly challenging. Short possessions compress already constrained activities into narrow periods. Set up and demobilisation consume a disproportionate share of available time. When delays occur, there is little opportunity to recover without carrying work forward into the next access period.

Over extended programmes, tunnel environments amplify fatigue and error risk. Work is physically demanding and cognitively taxing. Visibility is limited. Noise and heat are persistent. These conditions reduce tolerance for variation and increase reliance on procedural control. As pressure increases, the margin for safe improvisation narrows.

These realities have important implications for renewal strategy. Attempting to deliver major tunnel works through fragmented access can result in prolonged intermediate states that are difficult to stabilise. Temporary conditions persist. Interfaces remain open. The system becomes increasingly dependent on management control rather than on inherent robustness.

In such circumstances, the question is not whether tunnel renewals can be delivered within constrained access windows, but whether doing so represents the least risky approach for the system as a whole. This question becomes unavoidable as the renewal scope increases and as the physical limits of the environment are reached.

5.3. Viaduct and elevated structure constraints

Viaducts and elevated structures present a different but equally restrictive set of constraints for heavy civil renewal. While these assets are not enclosed like tunnels, their exposure introduces limitations that are often underestimated during planning.

Work on viaducts is governed by geometry, load paths and public interface. Available work space is limited. Edge protection, fall prevention and exclusion zones consume significant areas. Temporary works must be carefully designed to avoid overstressing the structure during intervention. In many cases, the structure itself becomes the governing constraint rather than the renewal activity being undertaken.

Access to viaducts is also highly dependent on logistics. Plant access, lifting strategy and material handling are constrained by height, reach and stability requirements. Opportunities for staging and storage are limited. Each intervention must be carefully sequenced to maintain structural integrity and ensure public safety beneath the railway.

Unlike tunnel environments, viaducts are exposed to weather and environmental conditions. Wind, temperature and precipitation can directly affect productivity and safety. Work that appears straightforward in principle may be repeatedly delayed or interrupted due to conditions beyond the control of the delivery organisation. Over time, these disruptions erode confidence in programme assumptions and increase delivery pressure.

The public interface beneath viaducts introduces further complexity. Roads, buildings and public spaces restrict access arrangements and emergency response. Noise, debris and visual intrusion are highly visible and politically sensitive. Possession overruns or unexpected disruption can have immediate consequences beyond the railway boundary.

As with tunnel renewals, attempts to fragment heavy civil work on viaducts into short access windows often result in prolonged intermediate states. Temporary works remain in place. Structural capacity may be partially reduced. Inspection and monitoring requirements increase. The system continues to operate, but with diminished tolerance for additional intervention (Figure 5.2).

Figure 5.2 Extended access decision threshold

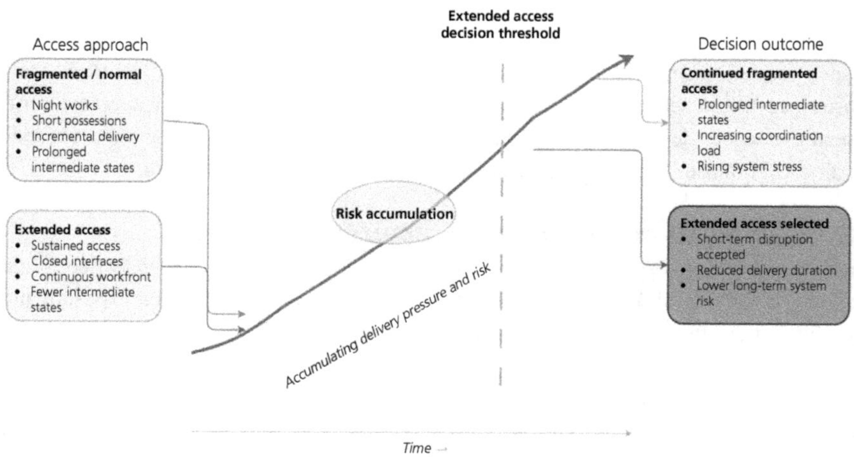

These conditions challenge the assumption that elevated structures offer greater flexibility for renewal. In practice, viaducts typically demand sustained access and tightly controlled working environments to manage risk effectively. Where this reality is not acknowledged, renewal strategies become driven by access availability rather than by engineering judgement.

Understanding viaduct constraints alongside tunnel environments reinforces a central theme of MRT renewal. The physical environment in which work is undertaken determines what is feasible, safe and sustainable. When access strategies do not reflect this reality, risk is displaced rather than reduced.

5.4. Slab track and drainage renewal challenges

Slab track and drainage systems represent some of the most challenging assets to renew within live MRT networks. Their complexity arises not from sophistication, but from embedment. These systems are integral to the structure and behaviour of the railway and are difficult to access, modify or replace without affecting surrounding assets.

Slab track is designed for stability and durability. Once installed, it becomes part of the structural system, transferring loads directly into the formation or structure beneath. Renewal therefore involves more than replacing track components. It requires breaking out concrete, managing reinforcement, controlling tolerances and reestablishing precise geometry. Each of these activities interacts with adjacent systems and demands sustained access.

Drainage systems present similar challenges. They are often out of sight and difficult to inspect directly. Over time, they become tightly coupled with surrounding ground conditions and structural elements. Renewal may require excavation, dewatering and reinstatement across extended lengths of the railway. In tunnel and viaduct environments, drainage works also interact with ventilation, electrical systems and structural durability.

Both slab track and drainage renewals resist fragmentation. Partial replacement can introduce new interfaces and transitional conditions that are difficult to manage. Leaving sections in intermediate states for extended periods increases dependence on temporary measures and monitoring. Over time, these arrangements can become the dominant source of risk rather than the original asset condition.

Access constraints magnify these challenges. Short possessions compress complex activities into narrow windows, increasing reliance on preparation and assumption. Set up and reinstatement consume a large proportion of available time. When issues arise, options for adjustment are limited. The work proceeds incrementally, but the system impact accumulates continuously.

In many MRT networks, these assets are renewed only once or twice within a generation. Organisational experience is limited. Lessons are learned slowly and often under pressure. The absence of repeatability makes reliance on routine delivery models particularly risky.

These characteristics explain why slab track and drainage renewals frequently act as inflection points in long-term programmes. They force a reassessment of access strategy, sequencing and risk tolerance. Attempting to deliver them as extensions of routine renewal activity often leads to prolonged disruption, escalating control measures and diminishing resilience.

Recognising the unique challenges of slab track and drainage renewal is essential for mature MRT asset management. It highlights the need to align access regimes with asset behaviour and to accept that some renewals require fundamentally different approaches to planning and delivery.

5.5. When continuous operation becomes unsustainable

The preceding sections illustrate a consistent pattern across heavy civil renewals. As work becomes more deeply embedded, more logistically constrained and more sensitive to intermediate conditions, the assumptions that support continuous operation come under increasing strain.

In the early stages of renewal, short access windows can often be made to work through careful planning, preparation and control. Over time, however, the cumulative effects of fragmentation become harder to manage. Temporary conditions persist. Interfaces remain open. Recovery margins erode. The system continues to operate, but with growing dependence on procedural control rather than inherent resilience.

At a certain point, the access model itself becomes the primary risk driver. The question facing decision makers is no longer how to deliver work within available windows, but whether those windows are compatible with the nature of the work being undertaken. This transition is rarely explicit. It is often reached gradually, as pressure accumulates and options narrow.

In tunnel and viaduct environments, this transition can arrive earlier than anticipated. Physical constraints, logistics and safety arrangements limit the extent to which work can be fragmented without introducing prolonged intermediate states. Slab track and drainage renewals further amplify these effects, resisting segmentation by design.

Sustained access or temporary suspension of service is often perceived as a last resort. In practice, it can represent a more controlled and predictable approach to managing risk. Longer access regimes allow work to be completed in stable conditions, interfaces to be closed decisively and systems to be returned to service in a known state. While disruption is more visible, uncertainty is reduced.

The reluctance to acknowledge this reality is understandable. MRT networks exist to provide continuity of service, and prolonged closures carry operational, political and reputational consequences. However, avoiding the question does not remove the underlying constraint.

It displaces risk into extended programmes, fatigued organisations and systems operating with diminished resilience.

Recognising when continuous operation becomes unsustainable is therefore a matter of engineering judgement and leadership. It requires an honest assessment of asset behaviour, environmental constraint and organisational capacity. Where this assessment is delayed, the eventual consequences are often more disruptive than a managed intervention would have been.

This recognition marks a turning point in mature MRT renewal programmes. It shifts the focus from preserving normal operation at all costs to sustaining the system over its remaining life. Chapter 6 examines how organisations can plan for disruption deliberately, rather than attempting to avoid it, and how access decisions can be governed in a way that supports long-term system integrity.

Managing Complexity in MRT Renewals: Systems, Interfaces, Governance and Leadership in Live Urban Rail Networks

Daniel Woods
ISBN 978-1-80791-579-7
https://doi.org/10.1108/978-1-80791-578-020261007

Chapter 6
Planning for disruption, not avoiding it

Disruption is an inherent feature of long-term MRT renewal. As networks age and renewal activity intensifies, the challenge is no longer whether service disruption will occur, but how it is anticipated, governed and absorbed by the system as a whole.

This chapter considers this condition as something to be planned for rather than avoided. It builds on the recognition in Chapter 5 that some renewal activities cannot be delivered safely or sustainably within normal access regimes. The focus therefore shifts from reactive mitigation towards deliberate planning, and towards managing disruption in a way that protects long-term system integrity.

By the time renewal programmes reach this stage, disruption is no longer a contingency to be managed. It becomes a strategic condition that must be deliberately designed, governed and absorbed by the system.

6.1. Disruption as a system condition

In many MRT organisations, disruption is treated as an exception. It is framed as a deviation from normal operation, something to be minimised through better planning or execution. While this mindset is understandable, it becomes increasingly misaligned with reality as renewal programmes mature.

Over extended renewal periods, service interruption becomes a recurring and predictable feature of system behaviour. Access constraints, heavy civil interventions and interface complexity combine to create conditions in which temporary loss of service or reduced capability is unavoidable. In this context, such impacts are not a failure of delivery. They are a consequence of sustaining an ageing system.

Treating disruption as an exception encourages reactive behaviour. Plans are optimised to preserve normal operation for as long as possible. Decisions are deferred. Temporary arrangements are extended. When interruption eventually occurs, it often does so abruptly and under pressure, with limited opportunity to manage its broader effects.

By contrast, treating this condition as a system property allows it to be engineered deliberately. Service reductions, access regimes and recovery strategies can be designed to align with renewal scope and system behaviour. Interfaces can be closed decisively rather than managed indefinitely. Organisational capacity can be planned around known periods of intensity rather than stretched continuously.

This shift requires a change in perspective. The objective is no longer to eliminate disruption, but to control its form, duration and impact. This includes recognising that some forms of intervention are less risky, less costly and less damaging to public trust than prolonged instability.

Service disruption also interacts with human and organisational factors. Extended periods of marginal operation place sustained demands on staff and decision makers. Fatigue accumulates. Tolerance for variation reduces. Planning for disruption explicitly allows these effects to be managed rather than endured.

Understanding disruption as a system condition is therefore foundational to mature MRT renewal. It provides the basis for making access decisions that reflect asset behaviour, environmental constraint and organisational capacity. Without this understanding, disruption remains something to be feared rather than something to be governed (Figure 6.1).

Figure 6.1 Disruption as a designed system condition

6.2. Designing disruption into renewal strategy

Once disruption is recognised as an inherent system condition, it must be addressed deliberately rather than reactively. This requires service impacts to be designed into renewal strategy from the outset, rather than treated as a contingency to be managed later.

In many MRT renewal programmes, these impacts are considered only after scope, schedule and access assumptions have been established. At that point, options are limited. Decisions are framed around mitigation rather than choice. This sequencing reinforces the belief that service disruption reflects a failure of execution rather than the consequence of strategic intent.

Designing disruption into renewal strategy reverses this logic. It starts by acknowledging the interaction between asset behaviour, environmental constraint and organisational capacity. Access regimes, service impacts and recovery strategies are considered alongside engineering scope, not downstream of it.

This approach requires explicit choices. It involves determining where sustained access, extended possessions or temporary suspension of service provides greater control over risk than fragmented delivery. It also requires clarity about which parts of the system can tolerate prolonged intermediate states and which cannot.

Importantly, designing disruption does not mean maximising service impact. It means shaping it. Shorter but more intense interventions may be preferable to prolonged periods of marginal operation. Clear and bounded disruption can be easier to manage than extended instability, both technically and organisationally.

This perspective also influences sequencing. Renewal activities that introduce significant interface change or require system integration are often better grouped within defined intervention periods rather than distributed across long programmes. Doing so allows interfaces to be closed decisively and assumptions to be reset in a controlled manner.

Designing disruption into strategy further supports realistic resource planning. Workforce availability, fatigue management and organisational capacity can be aligned with known periods of intensity. This reduces reliance on sustained overtime and informal adaptation that characterise many extended renewal programmes.

Ultimately, this approach reframes disruption as a tool rather than a threat. When disruption is planned, bounded and governed, it becomes a means of protecting long-term system integrity. Where it is avoided or deferred, risk tends to surface later in less predictable and harder to control.

6.3. Cost, capacity and the system response to disruption

MRT networks exist to move large numbers of people reliably through constrained urban environments. Any interruption to service therefore carries immediate consequences that extend beyond the engineering task itself. Loss of revenue, passenger displacement and wider network impacts all influence how renewal strategies are conceived and constrained.

The scale of these impacts is one of the primary reasons service disruptions are resisted. Reduced service affects fare income, but the wider cost is often felt through crowding on parallel routes, increased journey times and pressure on other transport modes. These effects are highly visible and politically sensitive, particularly in dense urban settings where alternatives are limited.

As a result, disruption is rarely considered in isolation. It is absorbed through system responses designed to preserve as much network capacity as possible. Replacement bus services, shuttle operations and partial line working are not simply mitigations. They are engineered operating modes that allow renewal to proceed while maintaining a degree of system functionality.

The feasibility of these responses depends heavily on network configuration. Linear routes with few crossover connections behave very differently from networks with multiple branches, turnbacks or loops. In some cases, sections can be isolated and operated independently. In others, a single blockage propagates disruption across large parts of the network. These structural characteristics shape what forms of disruption can be absorbed and for how long.

Staged renewal strategies often emerge from these constraints. Single line working, shuttle services between defined sections or temporary terminal arrangements can allow parts of the network to remain operational while work proceeds elsewhere. While such strategies can reduce passenger displacement, they introduce their own complexity. Operating patterns change. Recovery margins reduce. Interfaces multiply.

Replacement transport presents further challenges. Bus services rarely match rail capacity, particularly during peak periods. Routing, traffic conditions and crew availability all limit effectiveness. While replacement services are essential, they often serve to redistribute disruption rather than eliminate it. Their presence can create an illusion of equivalence that masks reduced system resilience.

These dynamics have important implications for renewal planning. Decisions about access and disruption cannot be separated from considerations of system capacity and demand. The question is not simply whether a section can be closed, but how the system responds when it is. In some configurations, short and intense disruption may be absorbed more

Figure 6.2 Cost, capacity and the system response to disruption

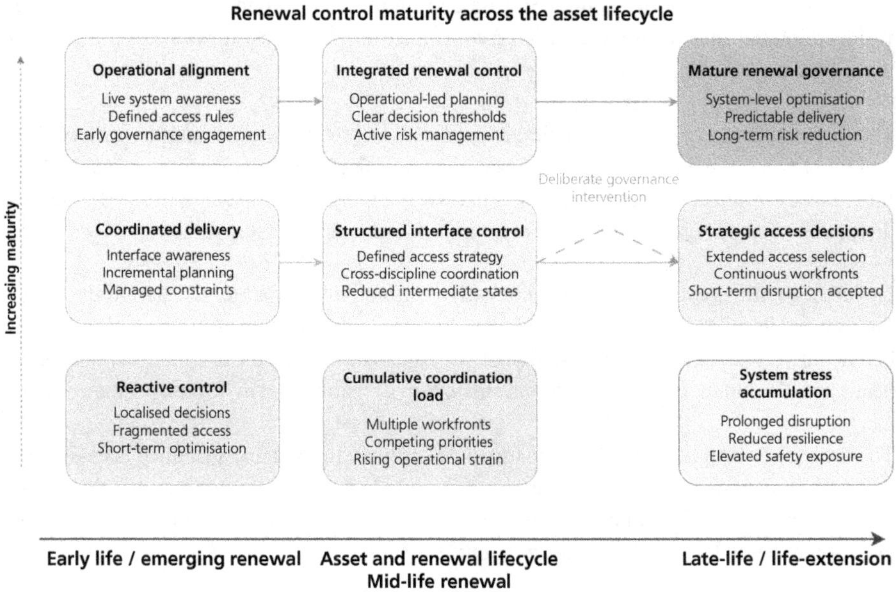

effectively than prolonged partial operation. In others, staged approaches may offer a more balanced outcome.

Understanding cost and capacity in this way reinforces the need to treat disruption as a system design problem. Engineering scope, access regimes and operating strategies must be considered together. Where this integration is absent, renewal decisions tend to be driven by immediate service preservation rather than by long-term system sustainability.

This perspective also clarifies why disruption decisions are inherently strategic. They involve trade-offs between visibility and duration, between short-term impact and long-term resilience. These competing pressures cannot be resolved through engineering analysis alone. They require governance structures capable of balancing technical, operational and societal considerations over extended periods (Figure 6.2).

6.4. Governing disruption decisions

Decisions to introduce sustained disruption, extended access or temporary suspension of service are among the most consequential choices made during MRT renewal programmes. They are also among the most difficult to govern.

These decisions typically sit at the intersection of engineering feasibility, operational consequence, public impact and organisational risk appetite. No single function owns all dimensions. As a result, authority is often diffuse and timing is uncertain. Decisions are deferred while additional analysis is undertaken or while alternative delivery approaches are explored within existing access regimes.

This deferral is rarely explicit. It emerges gradually as programmes attempt to preserve normal operation for as long as possible. Each incremental adaptation appears reasonable. Over time, however, the cumulative effect is to narrow the range of available options. By the time disruption becomes unavoidable, it is often introduced under pressure rather than by design.

Governance arrangements can unintentionally reinforce this dynamic. Escalation thresholds may be unclear. Decision forums may focus on delivery performance rather than system behaviour. Accountability for long-term outcomes may be separated from authority to approve near-term disruption. In these conditions, disruption decisions tend to drift rather than resolve.

The timing of disruption decisions is particularly critical. Introducing sustained access earlier in a programme may appear premature or excessive. Introducing it later may be framed as a response to failure. This framing influences organisational behaviour and public perception, even when the underlying engineering rationale is sound.

Effective governance does not eliminate this tension, but it makes it explicit. It requires a clear understanding of who has the authority to approve disruption, on what basis and at what point in the programme. It also requires decision makers to consider not only immediate service impacts, but the trajectory of risk over the remaining life of the renewal.

Governing disruption decisions therefore demands more than technical justification. It requires organisational maturity, trust between functions and a willingness to accept visible impact in order to protect long-term system integrity. Where this maturity is absent, disruption is often experienced as a failure rather than as a managed outcome.

This concludes the discussion on planning for disruption. Chapter 7 turns to the human dimension of these decisions, examining how judgement is exercised under pressure and why even well-governed organisations can struggle to act decisively when stakes are high.

Managing Complexity in MRT Renewals: Systems, Interfaces, Governance and Leadership in Live Urban Rail Networks

emerald PUBLISHING ice Publishing

Daniel Woods
ISBN 978-1-80791-579-7
https://doi.org/10.1108/978-1-80791-578-020261008

Chapter 7
Human factors and decision making under pressure

MRT renewals are delivered through engineering systems, but they are governed through human judgement. As access tightens, disruption increases and interfaces multiply, the quality of decision making becomes as important as the quality of technical design.

This chapter examines how decisions are made during MRT renewals when conditions are constrained, information is incomplete and consequences are significant. It focuses on the human factors that shape judgement, escalation and risk perception, and why these factors become more influential as renewal activity intensifies. Understanding these dynamics is essential to governing renewal programmes effectively over long periods.

7.1. Decision making in constrained environments

Decision making in MRT renewals rarely occurs in stable or neutral conditions. Choices are typically made within fixed access windows, under time pressure and with limited opportunity for recovery. Information is often incomplete, fragmented or late. The consequences of error are visible and immediate.

In these environments, individuals rely heavily on experience and pattern recognition. Formal analysis and assurance processes remain important, but they cannot always keep pace with events. Judgement bridges the gap. This reliance on judgement is not a weakness. It is an adaptive response to complexity.

However, renewal programmes systematically increase the demands placed on decision makers. Interfaces proliferate. Temporary arrangements persist. Assumptions change faster than documentation can be updated. Over time, this increases cognitive load and reduces the capacity for reflective decision making.

Pressure also reshapes how risk is perceived. Immediate delivery concerns tend to dominate attention. Risks that threaten short-term service continuity are prioritised over those that accumulate more slowly. Decisions that preserve stability today may be favoured over those that reduce longer term system risk.

Escalation behaviour is particularly affected. As abnormal conditions become familiar, thresholds for intervention tend to rise. Issues that would once have prompted challenge or pause are accepted as part of normal operation. This normalisation of strain allows programmes to continue, but it also masks the progressive erosion of resilience.

Organisational context reinforces these effects. Signals from leadership, performance measures and governance forums influence what is seen as acceptable. When success is defined narrowly, decision making adapts accordingly. Individuals act rationally within the system they inhabit, even when outcomes are undesirable.

Understanding decision making in constrained environments shifts attention away from individual error and towards the conditions that shape judgement. This perspective is critical for MRT renewals, where long-term outcomes are determined less by isolated decisions than by patterns of behaviour sustained over time. This is why technical controls alone are insufficient in long-duration renewal environments.

7.2. Experience and informal judgement

Experience plays a central role in how MRT renewals are delivered and governed. In complex and time constrained environments, experienced practitioners rely on informal judgement and pattern recognition to navigate situations that cannot be fully analysed in advance.

This form of judgement is not guesswork. It is built through repeated exposure to similar conditions and outcomes. Experienced engineers, operators and managers recognise familiar signals in system behaviour, interface stress and organisational response. They anticipate problems not because they have calculated every possibility, but because they have seen comparable situations unfold before.

Informal judgement often fills the space between formal processes. Standards, procedures and assurance requirements define acceptable boundaries, but they cannot account for every interaction or contingency. In practice, decisions are shaped by an understanding of how the system actually behaves, where flexibility exists and which risks are likely to grow if left unaddressed.

In long running renewal programmes, this experience-based judgement becomes particularly valuable. Conditions evolve gradually. Temporary arrangements persist. Small changes accumulate. Experienced individuals are often the first to sense when assumptions no longer hold or when the system is approaching a critical threshold. These signals are rarely precise, but they are often reliable.

However, reliance on experience also introduces vulnerability. Informal judgement is rooted in past conditions. When renewal activity alters the system in fundamental ways, previous patterns may no longer apply. New technologies, modified operating regimes or reconfigured interfaces can render earlier experience only partially relevant.

There is also a risk that critical judgement becomes concentrated in a small number of individuals. When experience is not shared or articulated, the system becomes dependent on personal knowledge. Organisational change, staff turnover or fatigue can then remove essential insight without warning.

Renewal programmes can unintentionally weaken experience-based decision making. Extended reliance on short-term resourcing, frequent reorganisation and fragmented delivery models disrupt continuity. Formal handovers rarely capture the subtle understanding of system behaviour that experienced practitioners carry. Over time, decision making becomes more procedural and less anticipatory.

Recognising the role of experience and informal judgement is therefore essential. It allows organisations to value professional judgement without over relying on it, and to understand where experience must be supported by continuity, challenge and shared understanding. Without this balance, MRT renewals risk becoming either overconfident or excessively constrained.

7.3. Fatigue, cognitive load and sustained renewal pressure

Long-term MRT renewal programmes place sustained demands on the people responsible for planning, delivery and operation. These demands extend beyond physical fatigue to include cognitive load, attentional strain and decision saturation.

Night work, extended access periods and prolonged periods of disruption disrupt normal work patterns. Recovery time is limited. Individuals operate across overlapping shifts and responsibilities. Over time, this erodes alertness and reduces the capacity to process complex information reliably.

Cognitive load increases as renewal progresses. Temporary arrangements accumulate. Interfaces multiply. Operating conditions deviate from normal. Decision makers must hold more variables in mind while margins for error shrink. Even experienced practitioners can struggle to maintain a coherent picture of system state under these conditions.

As fatigue and cognitive load increase, decision making becomes more conservative and more reactive. Individuals focus on maintaining immediate stability and avoiding visible failure. Longer term considerations are deferred. This is a rational response to pressure, but it can entrench conditions that increase underlying risk.

Communication is also affected. Under sustained pressure, information exchange becomes narrower and more transactional. Context is lost. Assumptions go unchallenged. Signals that something is changing in the system may be dismissed or normalised.

Importantly, these effects do not arise from lack of competence or commitment. They emerge from the structure of the renewal environment. Highly capable teams can continue to deliver for long periods under strain, which can mask the gradual erosion of resilience. By the time performance degrades visibly, recovery options are limited.

Fatigue and cognitive load also influence escalation behaviour. Thresholds rise as abnormal conditions become familiar. Issues that would once have prompted intervention are accepted as manageable. This gradual shift often goes unnoticed until a more significant event occurs.

Understanding these dynamics is essential for MRT renewals. It highlights that human performance cannot be separated from access regimes, disruption strategy and organisational design. Where sustained pressure is treated as normal, risk accumulates quietly. Where it is acknowledged and managed, the system retains greater capacity to respond to change.

7.4. Normalisation of degraded conditions

In long running MRT renewal programmes, degraded conditions can gradually become accepted as normal. What begins as a temporary deviation introduced to enable renewal activity evolves into an assumed operating state.

This process is rarely deliberate. It emerges as organisations adapt to sustained pressure and constrained access. Temporary rules remain in place. Workarounds persist. Reduced margins become familiar. Over time, these conditions lose their sense of exception.

As degraded conditions normalise, risk perception shifts. Issues that would once have triggered concern are reframed as manageable. Performance is judged relative to the current state rather than against original design intent or operating assumptions. The system continues to function, but with reduced resilience.

Normalisation is reinforced by organisational memory. New staff join programmes where degraded conditions are already established. Without experience of the system in its intended state, they accept current arrangements as a baseline. The gap between designed capability and actual operation becomes less visible.

Documentation and assurance processes can unintentionally support this shift. Temporary arrangements are formalised. Deviations are reapproved. Controls are added to manage symptoms rather than to restore underlying capability. Over time, the system appears stable, even as its tolerance for further disruption diminishes.

This dynamic is particularly dangerous in MRT renewals because it masks the accumulation of risk. The absence of immediate failure is interpreted as success. Warning signs are subtle and distributed. By the time intervention is considered, options are constrained and disruption is more severe.

Reversing normalisation is difficult. It requires a deliberate act of recognition and challenge. Assumptions must be revisited. Temporary conditions must be questioned. This often involves uncomfortable conversations about what has been accepted and why.

Understanding the normalisation of degraded conditions is essential for governing MRT renewals. It explains why problems are often recognised late and why decisive intervention can feel disproportionate to recent events. This understanding provides the foundation for addressing governance, accountability and assurance in the chapters that follow.

Human factors shape MRT renewal outcomes not through isolated errors, but through patterns of judgement exercised under sustained pressure. Experience, fatigue and adaptation interact with access constraints, disruption strategy and organisational context to influence how risk is perceived and managed over time.

Recognising these dynamics shifts attention away from individual performance and towards the conditions that shape decision making. This perspective is essential before examining governance, accountability and authority, where responsibility for managing renewal risk is formally assigned but often practically unclear. Chapter 8 turns to these structures directly, exploring how governance arrangements influence engineering judgement and system outcomes.

Managing Complexity in MRT Renewals: Systems, Interfaces, Governance and Leadership in Live Urban Rail Networks

emerald PUBLISHING ice Publishing

Daniel Woods
ISBN 978-1-80791-579-7
https://doi.org/10.1108/978-1-80791-578-020261009

Chapter 8
Governance, accountability and authority

MRT renewals are governed through formal structures that define responsibility, accountability and decision making authority. These structures are intended to provide clarity and control. In practice, they often struggle to keep pace with the complexity and uncertainty inherent in long-term renewal programmes.

This chapter examines how governance arrangements shape MRT renewal outcomes. It explores the alignment, or misalignment, between authority and risk, and why accountability can become diffuse in live and constrained environments. The chapter builds on earlier discussions of system behaviour, disruption and human decision making to show how governance influences what decisions are possible, when they are made and who is able to make them.

Once engineering limits are reached, governance becomes the primary constraint. How authority aligns with risk, and how decisions are escalated under pressure, increasingly determines renewal outcomes.

8.1. The alignment of authority and risk

In MRT renewals, risk rarely aligns neatly with organisational boundaries. Technical risk, operational consequence and public impact often sit with different parts of the organisation. Effective governance depends on how authority is distributed across these boundaries and how decisions are escalated when risk spans multiple domains.

In many renewal programmes, authority is fragmented. Engineering teams may identify emerging risk but lack the mandate to alter access regimes or operating strategy. Operational leaders may carry the consequences of disruption but have limited influence over renewal scope or sequencing. Senior governance forums may hold accountability without direct visibility of system behaviour.

This misalignment creates friction. Decisions are delayed while responsibility is clarified. Issues are managed locally rather than resolved systemically. Risk is tolerated rather than addressed because no single role is empowered to act decisively.

Governance frameworks often assume stability. Roles and responsibilities are defined on the basis of steady state operation or discrete projects. Long-term renewal challenges these assumptions. As conditions evolve, authority that was appropriate at one stage of the programme may no longer align with the risks being generated.

The most consequential governance failures in MRT renewals rarely involve the absence of structure. They involve structures that no longer match the problem being faced. Where authority does not follow risk, decision making becomes cautious and incremental, even as system pressure increases.

Aligning authority with risk does not require constant restructuring. It requires recognition that renewal programmes change the risk landscape and that governance arrangements must adapt accordingly. Without this recognition, even well designed governance systems can inhibit timely and effective decision making.

8.2. Accountability in long-term renewal programmes

Accountability in MRT renewals is difficult to sustain over long periods. Renewal programmes often extend well beyond typical organisational cycles, spanning multiple leadership teams, delivery models and strategic priorities. Over time, accountability can become diffused even when formal governance structures remain in place.

At the outset of renewal activity, accountability is usually clear. Roles are defined. Responsibilities are allocated. Reporting lines are established. As programmes progress, however, conditions change. Scope evolves. Access assumptions shift. Delivery strategies are adapted. Decisions made early in the programme continue to shape outcomes long after the individuals who made them have moved on.

This creates a form of temporal separation between decision and consequence. Accountability for long-term outcomes may rest with people who did not make the original choices, while those who did are no longer present to explain the context or intent. In such circumstances, accountability becomes procedural rather than substantive.

Organisational change can amplify this effect. Reorganisations, contract transitions and changes in delivery approach disrupt continuity. Accountability is reassigned, but understanding of system history is not always transferred with it. Over time, renewal programmes can lose a shared narrative of why certain decisions were made and what risks were consciously accepted.

In response, organisations often increase formal controls. Assurance processes are strengthened. Reporting becomes more detailed. While these measures provide visibility, they do not necessarily restore accountability. They can shift focus towards compliance with process rather than ownership of outcomes.

Figure 8.1 Governance layers and decision authority

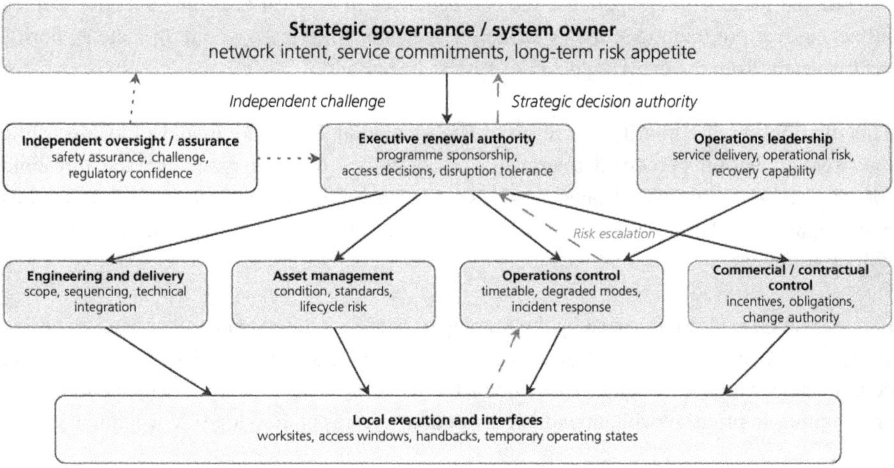

Accountability is further complicated by the distributed nature of MRT systems. Renewal risk often spans technical, operational and organisational domains. No single role experiences all consequences directly. When accountability is divided across functions, there is a tendency for each to manage its own exposure while assuming that system level risk is being addressed elsewhere.

Sustaining accountability in long-term renewal programmes therefore requires more than clearly defined roles. It requires continuity of intent, shared understanding of system risk and deliberate mechanisms to connect decisions made at different stages of the programme. Without these connections, accountability becomes fragmented and reactive.

Understanding how accountability evolves over time is critical for MRT renewals. It explains why programmes can appear well governed while still drifting towards increased risk. It also highlights the importance of aligning accountability with system outcomes rather than with organisational structure alone (Figure 8.1).

8.3. Authority to intervene and the problem of late escalation

In MRT renewal programmes, the recognition of emerging risk does not always translate into timely intervention. This gap is rarely due to a lack of awareness. More often, it reflects uncertainty about who has the authority to act and under what conditions escalation should occur.

Authority to intervene is often distributed across multiple layers of governance. Technical teams may identify concerns but lack the mandate to alter access regimes or delivery strategy. Operational leaders may experience the consequences of renewal decisions but have limited influence over programme scope or sequencing. Senior forums may hold ultimate authority, yet be distant from the evolving detail of system behaviour.

This distribution of authority can result in late escalation. Issues are managed locally while they remain within perceived tolerance. As conditions degrade gradually, the threshold for intervention shifts. What once would have prompted a challenge becomes accepted as manageable. Escalation is deferred in the hope that adaptation within existing constraints will be sufficient.

Late escalation is also influenced by the framing of decisions. Intervening often requires visible disruption, extended access or changes to agreed commitments. These actions carry immediate consequences and attract scrutiny. In contrast, continuing with existing arrangements allows programmes to progress without explicit disruption, even as underlying risk accumulates.

Governance structures often unintentionally reinforce this dynamic. Escalation pathways may be formal but ambiguous. Decision forums may focus on delivery progress rather than on leading indicators of system strain. Authority to pause, reset or reframe renewal strategy may exist in principle but be difficult to exercise in practice.

The result is a pattern in which decisive intervention occurs only after options have narrowed. By the time authority is exercised, disruption is more extensive and recovery more difficult. The intervention itself may then be perceived as disproportionate, despite being the outcome of long deferred decisions.

Addressing this problem requires recognising authority to intervene as a critical element of governance, not an exceptional measure. It requires clarity about who can initiate change, what information is needed to support that decision and how early warning signs are interpreted. Without this clarity, escalation becomes reactive and intervention arrives late.

Understanding the relationship between authority and escalation completes the governance picture for MRT renewals. It explains why risk is often managed incrementally rather than resolved decisively, and why renewal programmes can drift into constrained positions despite widespread professional competence.

Governance structures shape MRT renewal outcomes not through their formality, but through how authority, accountability and intervention are exercised over time. Where these elements are misaligned, risk tends to accumulate quietly, even in well intentioned and capable organisations.

Understanding governance in this way shifts attention away from compliance and towards system stewardship. It highlights the importance of continuity, clarity of authority and timely escalation in long-term renewal programmes. Chapter 9 examines how competence, succession and organisational memory influence the ability of MRT systems to sustain this stewardship across decades of renewal activity.

Managing Complexity in MRT Renewals: Systems, Interfaces, Governance and Leadership in Live Urban Rail Networks

Daniel Woods
ISBN 978-1-80791-579-7
https://doi.org/10.1108/978-1-80791-578-020261010

Chapter 9
Competence, succession and organisational memory

Long-term MRT renewal programmes depend not only on technical solutions and governance structures, but on the sustained availability of competence and judgement. Unlike discrete projects, renewal activity unfolds over decades, requiring organisations to retain and renew capability while the system itself is being transformed.

This chapter examines how competence, succession and organisational memory influence MRT renewal outcomes. It explores why capability is difficult to sustain over long periods, how experience is lost or diluted and why the erosion of organisational memory can become a critical source of risk in mature renewal programmes.

9.1. Competence as a system property

Competence in MRT renewals is often discussed as an attribute of individuals. In practice, it is a property of the system as a whole. It emerges from the interaction between people, processes, organisational structure and continuity of experience.

Individual expertise is essential, but it is not sufficient. Renewal programmes require collective understanding of system behaviour, interface sensitivity and historical context. This understanding cannot be fully captured in procedures or drawings. It develops through sustained involvement and shared experience over time.

In long running renewal programmes, competence must be renewed as assets are renewed. This creates a paradox. As experienced individuals move on or retire, the system loses knowledge at the same time as complexity increases. New staff inherit decisions, assumptions and constraints without always understanding their origins.

Formal competence frameworks provide structure, but they tend to focus on role definitions and technical qualifications. They are less effective at capturing judgement developed through exposure to system behaviour under pressure. As a result, organisations can appear well resourced while lacking depth of understanding in critical areas.

Delivery models can further complicate this picture. Reliance on short-term contracts, external delivery partners or rotating teams can fragment experience. Knowledge becomes distributed across organisational boundaries and is not always retained within the asset owner or operator. Over time, the system becomes dependent on individuals rather than on embedded capability.

Competence is also shaped by opportunity. Renewal programmes that progress incrementally over long periods may limit exposure to critical events. Individuals may work for years without experiencing full system integration, extended disruption or recovery from major intervention. When such events eventually occur, the organisation may find that competence exists in theory but not in practice.

Recognising competence as a system property shifts attention away from individual performance and towards organisational design. It highlights the importance of continuity, overlap and deliberate exposure to complexity. Without these elements, MRT renewal programmes risk becoming technically compliant but increasingly fragile.

9.2. Succession and the loss of experience

Succession in MRT renewal programmes is rarely a single event. It unfolds gradually as individuals rotate roles, move between organisations or leave the programme entirely. Over time, this gradual change can result in the loss of experience that is difficult to detect until its absence is felt.

In long-term renewal activity, experience is accumulated unevenly. Certain phases of work generate a deep understanding of system behaviour, interface sensitivity and organisational response under pressure. When individuals who have lived through these phases depart, much of that understanding leaves with them, even when formal handover processes are followed.

Succession planning often focuses on role coverage rather than experience continuity. Positions are filled. Responsibilities are transferred. However, the knowledge required to exercise judgement in complex and uncertain conditions does not transfer as easily. New incumbents may be technically capable but unfamiliar with the specific history and sensitivities of the system they inherit.

This challenge is amplified when succession occurs across organisational boundaries. Changes in delivery partners, contracting models or governance arrangements can disrupt continuity further. Experience becomes fragmented, with different parts of the story held by different organisations. Reconstructing this history later is difficult and often incomplete.

Time pressure reinforces these effects. Renewal programmes rarely pause to allow new teams to develop familiarity with system behaviour. Work continues. Decisions must be made. New staff learn while delivering, often without the benefit of context that earlier decisions assumed.

The loss of experience also affects escalation and challenge. Individuals who are new to a programme may be less confident in questioning established assumptions or practices. Without lived experience of previous inflection points, it is harder to recognise when the system is approaching another threshold.

Addressing succession in MRT renewals therefore requires more than staffing plans. It requires deliberate overlap, mentorship and opportunities for experience to be shared across generations of the programme. Without these mechanisms, renewal activity becomes progressively less informed by past experience, even as complexity increases.

9.3. Organisational memory and long-term risk

Organisational memory plays a critical role in the management of long-term MRT renewal programmes. It provides continuity of understanding across phases of work, leadership changes and shifting delivery models. Where this memory erodes, risk accumulates even when formal controls remain in place.

Much of the most valuable organisational memory in MRT renewals relates not to what was done, but to why it was done. Decisions about access strategy, sequencing, temporary arrangements and risk acceptance are often made in response to specific conditions at a particular point in time. When the context for these decisions is lost, later teams inherit outcomes without understanding intent.

Formal records capture only part of this history. Reports, drawings and approvals document decisions, but they rarely convey the assumptions, trade-offs and uncertainties that shaped them. Over time, these assumptions harden into constraints that are no longer questioned because their origin is unclear.

As renewal programmes progress, organisational memory can fragment. Different functions retain different parts of the story. Delivery teams remember execution challenges. Operations remember disruption. Governance forums remember commitments. Without deliberate integration, no single group retains a complete picture of system evolution.

Changes in organisational structure can accelerate this fragmentation. Reorganisation, contract transitions and outsourcing redistribute knowledge without necessarily transferring understanding. The system continues to function, but its collective memory becomes thinner and more dispersed.

The loss of organisational memory has practical consequences. Warning signs are harder to interpret. Past inflection points are forgotten. The reasons for earlier interventions are no longer visible. As a result, the system may repeat patterns that were previously recognised as risky, without realising that they have been encountered before.

Preserving organisational memory in MRT renewals therefore requires active effort. It involves maintaining continuity across programme phases, creating opportunities for shared reflection and ensuring that critical context is carried forward as roles and teams change. Without this effort, renewal programmes risk becoming less informed over time, even as experience accumulates in isolated pockets.

The loss of organisational memory also weakens interface management, as assumptions embedded in earlier decisions become harder to recognise or challenge.

9.4. Role transition and capability shift

Long-term MRT renewal programmes often coincide with significant career progression for the individuals involved. As programmes extend over many years, people move into roles of greater authority, broader responsibility and increased visibility. These transitions are a natural and positive feature of organisational development, but they also change the demands placed on individuals.

Longevity within an organisation provides valuable continuity and familiarity with system context. However, the capability required to manage complex renewal activity differs from that required to deliver technical work or supervise discrete tasks. As roles evolve, the balance of skills shifts from technical execution towards judgement under uncertainty, integration across interfaces and ownership of system wide risk.

A similar transition occurs for technical specialists who move into management or leadership positions. Deep technical expertise remains valuable, but it is no longer sufficient on its own. Renewal leadership requires the ability to synthesise incomplete information, align diverse interests and make decisions that trade short-term impact against long-term system integrity. These demands are often unfamiliar and not explicitly supported.

Organisations do not always recognise this shift clearly. Promotion and role assignment may be based on experience, tenure or past performance in technical roles. While these attributes are important, they do not automatically prepare individuals for the different forms of responsibility that renewal management entails. Without deliberate support, individuals are expected to adapt while continuing to deliver under pressure.

This dynamic can create subtle risk. Individuals may remain closely involved in technical detail because it is familiar and tangible, while broader system issues receive less attention. Alternatively, decision making may become overly cautious as new responsibilities and visibility increase. Both responses are understandable, but they can affect how renewal risk is managed over time.

Recognising role transition as a capability shift rather than a simple change in title is therefore essential. It highlights the need for deliberate preparation, mentoring and shared decision making as responsibilities expand. In long-term MRT renewals, sustaining capability depends not only on retaining people, but on supporting them as the nature of their contribution evolves.

Competence in MRT renewals is sustained not through individual excellence alone, but through continuity of experience, shared understanding and deliberate support as roles evolve. Succession, organisational memory and role transition each shape how judgement is exercised over time, particularly under sustained pressure.

Recognising these factors shifts attention away from resourcing numbers and formal structures towards the conditions that allow capability to endure across decades of renewal. This perspective is essential before considering assurance, oversight and independent challenge, which are often relied upon to compensate for gaps in experience and organisational memory. Chapter 10 examines how these mechanisms function in complex renewal environments and where their limits lie.

Daniel Woods
ISBN 978-1-80791-579-7
https://doi.org/10.1108/978-1-80791-578-020261011

Chapter 10
Assurance, oversight and independent challenge

Assurance and oversight play a central role in MRT renewal programmes, particularly where risk is high, access is constrained and decisions carry significant public consequences. These functions are intended to provide confidence that renewal activity is being planned and delivered safely, competently and in the public interest. There is a critical distinction between assurance that builds confidence and assurance that attempts to enforce control.

This chapter examines how assurance, oversight and independent challenge operate in long-term MRT renewals. It explores the value these functions provide, the limitations they face in complex environments and the conditions under which they support sound engineering judgement rather than inadvertently constraining it. Understanding this balance is essential for governing renewal risk effectively over extended periods.

10.1. The purpose of assurance in MRT renewals

In its simplest form, assurance exists to provide confidence. In MRT renewals, that confidence must extend beyond technical compliance to include system behaviour, integration risk and the cumulative effects of long-term change.

Assurance functions are often framed as mechanisms for checking that requirements have been met. While this is important, it is insufficient on its own. Renewal programmes involve evolving scope, temporary arrangements and changing assumptions. The most significant risks arise not from noncompliance, but from interactions between elements that are individually compliant yet collectively unstable.

Effective assurance therefore focuses on understanding how the system is changing over time. It examines whether assumptions remain valid, whether interfaces are being managed coherently and whether emerging patterns indicate increasing strain. This requires assurance to engage with context and intent, not just documentation.

In long running renewal programmes, assurance also plays a stabilising role. It provides continuity as teams change, leadership rotates and delivery models evolve. Independent perspective helps surface issues that may no longer be visible to those embedded in daily delivery.

However, assurance operates within constraints. It relies on information provided by the programme. It is subject to governance structures and reporting cycles. Its influence depends on how its findings are received and acted upon. Where assurance is treated as a compliance hurdle, its ability to support system level understanding is limited.

The timing of assurance intervention is particularly important. Early engagement allows challenge to shape strategy and access decisions while options remain open. Late assurance tends to focus on justification of decisions already made, reducing its capacity to influence outcomes.

Understanding the purpose of assurance in this broader sense reframes its role. Assurance is not an external check applied to a finished plan. It is a mechanism for supporting judgement, surfacing uncertainty and maintaining visibility of system behaviour as renewal progresses.

10.2. Independent challenge and the value of distance

Independent challenge is a defining feature of effective assurance in MRT renewals. Its value lies not in technical superiority, but in distance from delivery pressure, organisational incentives and local norms that shape decision making over time.

As renewal programmes progress, those directly involved become embedded in the system they are managing. This embeddedness is unavoidable and often beneficial. It enables rapid decision making, informal coordination and adaptation to emerging issues. However, it also narrows the perspective. Assumptions become familiar. Temporary arrangements persist. Boundaries shift without being consciously revisited.

Independent reviewers and oversight bodies bring a different vantage point. They are able to question why certain approaches have been adopted, whether original assumptions still hold and how cumulative changes are affecting system behaviour. Their challenge is most valuable when it addresses intent and trajectory rather than isolated technical detail.

Distance also enables comparison. Independent challenge draws on experience across multiple programmes, networks and contexts. This comparative perspective allows patterns to be recognised that may not be visible within a single organisation. Similar issues arising in different systems can be identified as structural rather than incidental.

For independent challenge to be effective, it must be welcomed rather than tolerated. Where reviewers are positioned as obstacles to progress, engagement becomes transactional.

Findings are managed rather than explored. In these conditions, independence is preserved in form but not in influence.

Timing and scope are critical. Independent challenge that occurs too late is constrained by decisions already made and commitments already given. Early engagement allows challenge to shape strategy, access regimes and governance arrangements while alternatives remain available.

The credibility of independent challenge also depends on its understanding of context. Effective reviewers recognise the realities of live operation, political constraint and organisational pressure. Challenge that is detached from these realities risks being dismissed, regardless of technical merit.

Independent challenge therefore functions best as a partner in system stewardship. Its role is not to replace decision making, but to illuminate blind spots, surface emerging risk and support informed judgement. Where this relationship is established, assurance adds value beyond compliance and becomes an integral part of managing renewal complexity.

10.3. Oversight, reporting and the illusion of control

Oversight in MRT renewal programmes is commonly exercised through structured reporting, formal reviews and performance metrics. These mechanisms are intended to provide visibility, assurance and control in environments where risk is high and consequences are significant.

In complex renewal programmes, however, the appearance of control can diverge from actual system behaviour. Reporting frameworks tend to simplify reality. They stabilise narratives, aggregate information and filter uncertainty. While this makes oversight manageable, it can also mask emerging risk.

As renewal activity intensifies, the system becomes increasingly dynamic. Temporary arrangements proliferate. Interfaces shift. Local adaptations are made to maintain progress. Many of these changes are rational and necessary, but they are difficult to capture meaningfully in standard reporting formats.

Performance indicators often reinforce this effect. Metrics are typically aligned with delivery milestones, compliance status and immediate service impact. These measures provide useful signals, but they can draw attention away from leading indicators of strain such as increasing workarounds, rising cognitive load or narrowing recovery margins.

Oversight forums may therefore receive a picture of stability while underlying resilience is eroding. The absence of reported issues is interpreted as success. Gradual degradation remains invisible because it does not trigger thresholds designed for discrete events.

This dynamic can create an illusion of control. Confidence is derived from the regularity of reporting and the completeness of documentation rather than from direct understanding of system behaviour. Decision makers may feel well informed while remaining disconnected from how risk is evolving on the ground.

The challenge is compounded by the pressure to demonstrate progress. Renewal programmes operate under scrutiny from multiple stakeholders. Reporting becomes a means of reassurance as much as of insight. In such conditions, uncomfortable signals are more likely to be reframed than escalated.

Recognising the limits of oversight does not diminish its importance. It highlights the need for reporting that supports judgement rather than replaces it. Effective oversight complements structured information with narrative, challenge and direct engagement with the system. Without this balance, assurance can become an exercise in managing perception rather than managing risk.

10.4. When assurance constrains decision making

While assurance and oversight are essential in MRT renewals, they can also introduce constraints that influence how decisions are made. These constraints are rarely intentional. They emerge from the interaction between assurance processes, governance expectations and organisational risk appetite.

In highly scrutinised environments, assurance findings can carry significant weight. Decisions that deviate from established plans, introduce visible disruption or rely on professional judgement rather than documented precedent may attract challenge. Over time, this can shape behaviour. Decision makers may favour options that are easier to justify through existing assurance frameworks, even if alternative approaches offer better long-term outcomes.

This dynamic can be particularly pronounced when assurance is engaged late. Where strategic decisions have already been made, assurance activity may focus on validating compliance rather than exploring whether the strategy itself remains appropriate. In such cases, challenge is applied to execution rather than to intent.

There is also a risk that assurance becomes a proxy for decision making. When uncertainty is high, organisations may seek assurance endorsement as a way of transferring responsibility. This can slow response, dilute ownership and discourage timely intervention. Decisions that require judgement under uncertainty are deferred in favour of additional review.

The cumulative effect can be a narrowing of perceived options. Renewal strategies become constrained by what can be readily assured rather than by what the system requires. This does not eliminate risk. It redistributes it, often into longer term system behaviour that is harder to observe and harder to correct.

Avoiding this outcome does not require reducing assurance. It requires clarity about its role. Assurance should inform and challenge decision making, not replace it. It should create space for professional judgement rather than constrain it through excessive formality.

In mature MRT renewal programmes, the relationship between assurance and decision making is explicit and balanced. Assurance provides independent insight and challenge, while accountability for decisions remains clearly held. Where this balance is achieved, assurance strengthens system stewardship rather than limiting it.

Assurance, oversight and independent challenge are critical elements of MRT renewal governance, but their effectiveness depends on how they are applied and integrated. When focused on system behaviour, timing and intent, assurance supports sound judgement and early intervention. When treated as a compliance function alone, it risks reinforcing inertia and delaying necessary change.

Understanding the role and limits of assurance completes the discussion of governance and capability in MRT renewals. The final chapters turn to leadership and long-term system maturity, examining how organisations can sustain renewal activity while preserving resilience, trust and public confidence over decades of change.

Managing Complexity in MRT Renewals: Systems, Interfaces, Governance and Leadership in Live Urban Rail Networks

Daniel Woods
ISBN 978-1-80791-579-7
https://doi.org/10.1108/978-1-80791-578-020261012

Chapter 11
Leadership in long-term MRT renewal programmes

Leadership in MRT renewals differs fundamentally from leadership in discrete projects or short-term programmes. The scale, duration and visibility of renewal activity place sustained demands on those responsible for setting direction, making trade-offs and maintaining organisational coherence over long periods of change.

This chapter examines leadership as a system function rather than a personal attribute. It explores how leaders influence renewal outcomes through the conditions they create, the decisions they prioritise and the behaviours they reinforce over time. Understanding leadership in this context is essential for sustaining renewal activity without eroding system resilience or public trust.

11.1. Leadership as system stewardship

In long-term MRT renewal programmes, leadership is exercised less through direct control and more through stewardship of the system as a whole. Leaders shape outcomes by defining intent, establishing boundaries and maintaining focus on long-term integrity amid competing pressures.

Stewardship requires an understanding of how decisions made today affect system behaviour years later. Access strategies, disruption tolerance and investment priorities all carry consequences that extend beyond individual programme phases. Leaders must therefore hold a longer time horizon than those responsible for day-to-day delivery.

This perspective challenges conventional leadership models that emphasise decisiveness and short-term results. In renewal environments, effective leadership often involves restraint. It requires resisting pressure to optimise individual interventions at the expense of cumulative system risk. It also requires recognising when stability is being preserved artificially rather than sustainably.

Leadership as stewardship is particularly important where renewal activity intersects with live operation. Decisions that maintain service continuity in the short term may increase fragility over time. Leaders must be willing to confront this tension openly and to support strategies that protect long-term system performance, even when immediate impacts are visible.

The authority of leaders in this context derives not only from position, but from credibility. Consistent messaging, alignment between words and actions and willingness to engage with uncomfortable constrained choices all influence how decisions are interpreted and implemented across the organisation.

Viewing leadership as system stewardship reframes accountability. Leaders are not solely responsible for outcomes, but they are responsible for the conditions under which decisions are made. Where these conditions encourage openness, challenge and strategic thinking, renewal programmes are better able to navigate complexity without drifting into reactive behaviour.

11.2. Leadership, trade-offs and decision timing

Leadership in MRT renewal programmes is exercised most visibly through the trade-offs that are accepted and the timing of decisions. Few choices are unambiguously correct. Most involve balancing competing objectives under conditions of uncertainty, constraint and scrutiny.

Decisions about access, disruption, scope and sequencing often present leaders with imperfect options. Preserving service continuity may increase delivery pressure. Accelerating renewal may heighten public impact. Deferring intervention may protect short-term stability while increasing long-term risk. Leadership judgement lies in recognising these tensions and acting deliberately rather than reactively.

Timing is critical. Decisions taken early can appear cautious or disruptive, particularly when problems have not yet materialised. Decisions taken later are often framed as responses to failure, even when the underlying issues were foreseeable. Leaders must therefore operate with an awareness of how decisions will be interpreted as well as how they will affect system behaviour.

Trade-offs are also shaped by organisational signals. Performance measures, reporting expectations and public messaging influence what types of decisions feel permissible. Where leaders emphasise short-term outcomes, decision making adapts accordingly. Where leaders reinforce strategic intent and system integrity, greater space exists for proactive intervention.

Effective leadership does not eliminate uncertainty. It acknowledges it openly. Leaders who are prepared to articulate the rationale behind difficult decisions create conditions in which decisions can be understood and supported, even when impacts are visible and uncomfortable.

Figure 11.1 Leadership influence over time

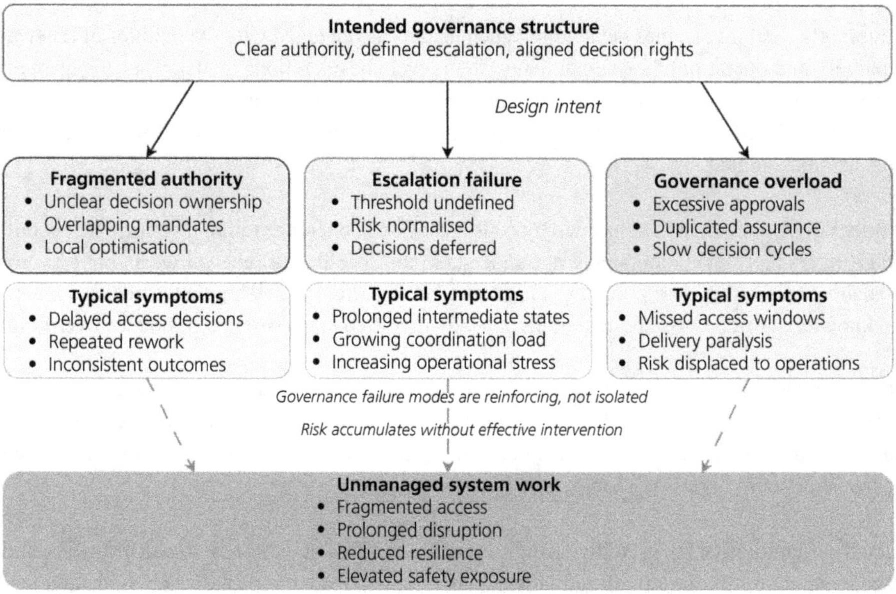

In MRT renewals, the accumulation of small decisions often matters more than any single intervention. Leadership is therefore expressed through consistency. Repeated choices that align with stated priorities reinforce credibility and shape behaviour across the organisation.

Leadership that engages deliberately with trade-offs and decision timing helps renewal programmes avoid drift. It enables complexity to be navigated consciously rather than absorbed incrementally, preserving both system resilience and organisational trust (Figure 11.1).

11.3. Leadership visibility, trust and public scrutiny

MRT renewal programmes operate under sustained public and organisational scrutiny. Service disruption, extended works and visible infrastructure intervention attract attention from passengers, media and political stakeholders. In this environment, leadership visibility takes on particular importance.

Visibility does not require constant presence or public commentary. It requires clarity about who is accountable for decisions and confidence that those decisions are being made deliberately. Where leadership is perceived as distant or reactive, uncertainty fills the gap. Trust is eroded not by disruption itself, but by the absence of clear ownership.

Internally, visibility shapes behaviour. When leaders are engaged with the realities of renewal delivery and operation, they signal that system level considerations matter. This engagement creates space for challenge and escalation. Conversely, when leadership appears disconnected, individuals are more likely to manage risk locally and quietly, allowing issues to persist unexamined.

Externally, public trust is influenced by consistency and coherence rather than technical detail. Passengers and stakeholders may not understand the specifics of renewal work, but they are sensitive to whether disruption appears planned or chaotic. Leadership that articulates intent, acknowledges trade-offs and maintains a steady narrative helps sustain confidence over time.

Public scrutiny also affects decision making indirectly. The anticipation of reaction can discourage early intervention and encourage incremental adaptation. Leaders who recognise this dynamic are better placed to support decisions that protect long-term system integrity, even when short-term impact is visible.

Trust is cumulative. It is built through repeated alignment between stated priorities and observed action. In long running renewal programmes, leadership credibility is shaped less by individual announcements and more by the consistency with which difficult decisions are approached.

Leadership visibility, when exercised with purpose and restraint, reinforces system stewardship. It enables organisations to navigate renewal under scrutiny without becoming reactive, preserving both operational stability and public confidence.

11.4. Leadership judgement under competing pressures

Leadership judgement in MRT renewal programmes is exercised under conditions where pressures rarely align. Operational continuity, safety assurance, political expectation and financial constraint exert influence simultaneously. Few decisions optimise all dimensions. Most require acceptance of imbalance.

Under these conditions, judgement is not about finding the correct answer. It is about selecting a course of action that the system can absorb and sustain. Leaders must consider not only whether a decision is technically defensible, but whether it can be implemented, communicated and maintained over time.

Competing pressures often present as urgency. Service reliability, public reaction and programme milestones create momentum that favours immediate action. Strategic judgement requires the ability to slow decision making when necessary, to test assumptions and to consider secondary effects that may not be immediately visible.

Judgement is also shaped by organisational context. Leaders operate within governance arrangements, assurance expectations and performance frameworks that influence what feels permissible. Recognising these influences allows leaders to distinguish between genuine constraint and perceived limitation.

In renewal environments, judgement is exercised repeatedly rather than episodically. Small decisions accumulate. The consistency of these decisions matters. Leaders who apply judgement unevenly or shift priorities frequently create uncertainty and invite reactive behaviour across the organisation.

Effective leadership judgement balances resolve with openness. It accepts responsibility for difficult choices while remaining receptive to challenge and emerging information. This balance enables leaders to adjust course without undermining credibility.

Judgement under competing pressures is therefore not a personal trait but a practice. It develops through experience, reflection and exposure to consequence. In MRT renewals, where decisions echo across years of operation, leadership judgement shapes not only project outcomes but system behaviour over the long term.

Leadership in MRT renewal programmes is defined less by position than by the quality of judgement exercised over time. Through stewardship, deliberate trade-offs and consistent decision making, leaders shape the conditions under which renewal activity unfolds and risk is managed.

Understanding leadership in this way highlights its cumulative impact on system behaviour. It also underscores the importance of aligning intent, action and communication across long periods of change. The final chapter draws together the themes of this book, examining how MRT organisations can move towards more mature renewal systems that balance continuity, resilience and public trust.

Managing Complexity in MRT Renewals: Systems, Interfaces, Governance and Leadership in Live Urban Rail Networks

Daniel Woods
ISBN 978-1-80791-579-7
https://doi.org/10.1108/978-1-80791-578-020261013

Chapter 12
Towards mature MRT renewal systems

MRT renewals are not defined by any single project, intervention or decision. They unfold through a series of choices made over many years, shaped by system constraints, organisational capability and leadership judgement. As renewal activity becomes a permanent condition rather than an exceptional event, the maturity of the renewal system itself becomes the determining factor in long-term outcomes.

This chapter draws together the themes explored throughout the book and reflects on what characterises mature MRT renewal systems. It does not propose an endpoint or ideal state. Instead, it identifies recurring patterns that distinguish renewal environments able to adapt, learn and sustain performance under continuous change.

12.1. Renewal as a permanent system condition

In mature MRT networks, renewal is no longer a temporary phase. Assets age continuously. Technology evolves. Demand patterns shift. Renewal activity becomes embedded within normal operation rather than occurring as discrete episodes.

Recognising renewal as a permanent system condition changes how it is governed. Planning horizons extend beyond individual programmes. Access strategies are considered in the context of long-term system behaviour. Disruption is managed as an expected outcome rather than an exception.

Organisations that treat renewal as episodic tend to optimise locally and defer difficult decisions. Those that accept its permanence are more likely to invest in capability, continuity and system understanding. This shift in perspective underpins renewal maturity.

12.2. From project thinking to system stewardship

A central theme of this book has been the limitation of project-based thinking in MRT renewals. While projects provide structure and accountability, they are insufficient on their own to manage system level risk over decades.

Figure 12.1 Characteristics of mature MRT renewal systems

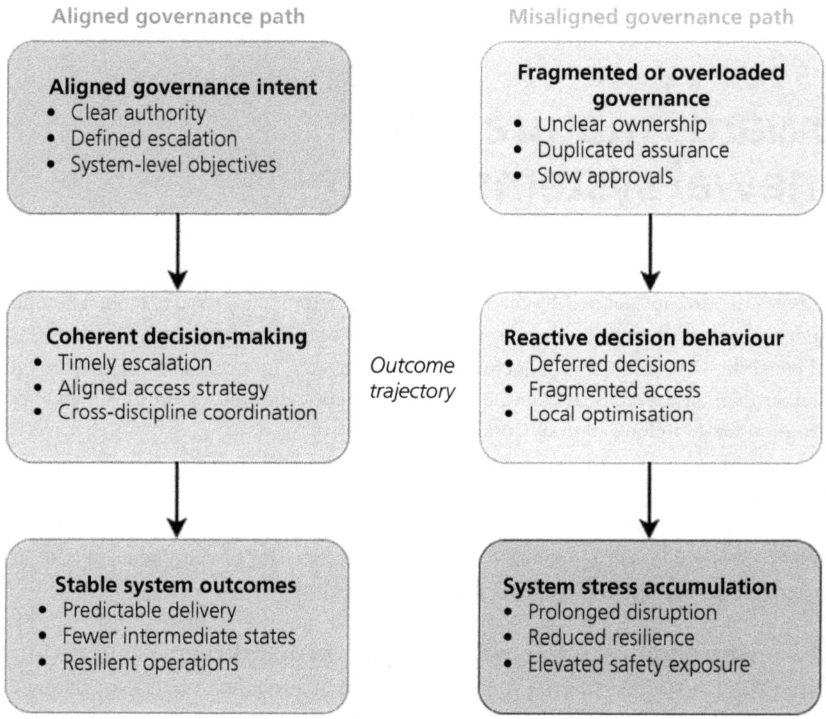

Mature renewal systems balance project delivery with system stewardship. Decisions are evaluated not only on their immediate outcomes, but on how they influence interfaces, organisational behaviour and future options. Success is measured in terms of resilience and adaptability as well as delivery performance.

This balance requires organisations to tolerate ambiguity. Not all benefits are immediately visible. Some decisions protect the system quietly by preserving flexibility and margin rather than by delivering tangible outputs. Recognising and valuing these contributions is a mark of maturity (Figure 12.1).

12.3. Capability, memory and the ability to learn

Renewal maturity depends heavily on the ability to retain and refresh capability over time. Competence, organisational memory and role transition shape how judgement is exercised as conditions evolve.

Mature systems create mechanisms for learning that extend beyond individual programmes. Experience is shared across phases of work. Assumptions are revisited deliberately. Past decisions are understood in context rather than treated as fixed constraints.

This capacity to learn is particularly important where renewal involves disruption and public impact. Organisations that learn effectively are better able to adjust strategy before pressure escalates, reducing the need for reactive intervention.

12.4. Governance that enables judgement

Governance in mature MRT renewal systems is characterised by clarity and adaptability. Authority aligns with risk. Accountability persists across organisational change. Assurance supports decision making without constraining it.

Rather than seeking certainty, mature governance frameworks create space for judgement under uncertainty. They encourage early escalation, independent challenge and deliberate trade-offs. Decisions are owned clearly, even when outcomes are uncertain.

This approach recognises that renewal risk cannot be eliminated. It can only be managed through informed choice and sustained oversight.

12.5. Leadership and long-term trust

Leadership remains central to renewal maturity. Leaders shape outcomes through the conditions they establish, the priorities they reinforce and the consistency of their judgement over time.

Trust is built gradually. Internally, it emerges from alignment between intent and action. Externally, it is sustained through transparency, coherence and visible ownership of difficult decisions. Mature MRT renewal systems are distinguished less by technical sophistication than by their ability to align engineering, governance and leadership judgement over time.

Leadership that acknowledges trade-offs openly and supports long-term system integrity strengthens this trust, even in the presence of sustained change.

Closing reflections

Managing complexity in MRT renewals requires more than technical expertise or procedural control. It demands an understanding of system behaviour, an appreciation of human and organisational dynamics and a willingness to engage with uncertainty deliberately.

Mature MRT renewal systems are not defined by the absence of disruption or risk. They are defined by how risk is recognised, governed and absorbed over time. By treating renewal as a permanent system condition and by aligning capability, governance and leadership accordingly, organisations can sustain safe and reliable operation while navigating decades of change.

This book has sought to articulate the patterns that shape renewal outcomes in live MRT environments. Its intent is not to prescribe solutions, but to support informed judgement in contexts where consequences are real and options are constrained. In doing so, it aims to contribute to the ongoing development of MRT systems that are resilient, trusted and capable of renewal for the long term.

Managing Complexity in MRT Renewals: Systems, Interfaces, Governance and Leadership in Live Urban Rail Networks

Daniel Woods

ISBN 978-1-80791-579-7
https://doi.org/10.1108/978-1-80791-578-020261014

Index

Note: Italics in page numbers indicates figures.

www.ingramcontent.com/pod-product-compliance
Lightning Source LLC
LaVergne TN
LVHW021313070826
845179LV00004B/35